Stripping Down Science

DR CHRISTOPHER SMITH

Stripping Down Science

THE NAKED SCIENTIST BARES THE FACTS

WILLIAM HEINEMANN: AUSTRALIA

A William Heinemann book
Published by Random House Australia Pty Ltd
Level 3, 100 Pacific Highway, North Sydney NSW 2060
www.randomhouse.com.au

First published by William Heinemann in 2010

Addresses for companies within the Random House Group can be found at
www.randomhouse.com.au/offices.

National Library of Australia
Cataloguing-in-Publication Entry

Smith, Christopher, 1975–.
Stripping down science: the naked scientist exposes the facts.

ISBN 978 1 74166 645 8 (pbk).

Science – Popular works.

Dewey classification: 500

Cover design by Luke Causby/Blue Cork
Internal design and typesetting by Xou Creative, www.xou.com.au
Internal illustrations by Shane Nagle, www.shanenagle.com
Printed and bound by Griffin Press, an accredited ISO AS/NZS 14001:2004
Environmental Management System printer

10 9 8 7 6 5 4 3 2 1

The paper this book is printed on is certified against
the Forest Stewardship Council® Standards. Griffin
Press holds FSC chain of custody certification
SGS-COC-005088. FSC promotes environmentally
responsible, socially beneficial and economically viable
management of the world's forests.

To Sarah, Amelia and Tim,
whom I adore.

Contents

Age no impediment to hot sex!

Most people are comfortable with the concept that plants use sweet rewards to attract insects, which fly in for a drink of nectar and pick up pollen in the process. When the same bugs later drop in on other plants of the same species, they shed some of the pollen and fertilise the flowers.

But it's not true that plants aim to be attractive to insects all the time, as scientist Irene Terry from the University of Utah, together with colleagues at the University of Queensland,* discovered when she began to study one member of a family of primitive plants called cycads. Dubbed 'living fossils' because they have hardly changed since the time of the dinosaurs, cycads resemble palms or ferns, although they are not related to either. In fact, they're members of the conifer and fir family and, consequently, reproduce using a fir-cone-like fruiting body which pops up for a fertile period lasting about four weeks every one-to-four years.

Like humans, these plants come in male and

* *Science* 5 October 2007: Vol. 318. no. 5847, p. 70 DOI: 10.1126/science.1145147

female varieties, with the male cones producing pollen, which then fertilises the female cones. Previously, scientists thought that the wind carried the pollen between the two, until it was realised that the scales that make up the cones are packed together too tightly for the pollen to enter efficiently. Intrigued by the mystery, the researchers subsequently discovered that a species of insect called a thrip – in this case *Cycadothrips chadwicki* – was responsible. The insects visit the male cones to eat the pollen and then carry some of it to the female cones. But therein lies a problem: how to persuade the insects to abandon their banquet in the male cones and visit the female cones, which are devoid of pollen and therefore food?

It turns out that the cycads resort to the plant equivalent of chemical crowd control to force the insects to pay for their free lunch. Each day, between 11 am and 3 pm, the temperature of the cones, and particularly the male cones, shoots up by over 12 degrees Celsius. The plants achieve this thermal feat by breaking down stockpiled starch, sugars and fats. The resulting temperature boost makes the environment inside the cone uncomfortable for the thrips. It's also accompanied

by a million-fold surge in the production of an odorous chemical called beta-myrcene.

At low levels, thrips find the smell attractive, which helps to lure them into the cone in the first place; but at higher concentrations it smells repugnant and drives them out. Humans also find the stink offensive. 'It takes your breath away. It's a harsh, overwhelming odour like nothing you ever smelled before,' remarks Irene Terry, presumably while holding her nose.

So, repelled by the smell but still covered in the vestiges of their last pollen meal, the insects abandon their feast and seek refuge in the surrounding bushes. By mid-afternoon the cones have cooled again and the concentration of beta-myrcene drops, encouraging the thrips to return, pollen still clinging to their bodies. And, since the male and female cones both look and smell alike, the insects flock back to both, carrying pollen into the female cones and fertilising them.

The cycad cycle repeats itself day after day 'until the males wear out and the females are happily pollinated', says Terry. Not that different from humans then, despite the 300-million-year age gap!

Sexual subversion elsewhere in nature . . .

It's not just cycads that have mastered the art of olfactory allure. Other species have also evolved to take advantage of insects' insatiable desires for food, including an orchid that makes itself pong like a bee to attract a hungry hornet.

Until recently, no one was sure what pollinated the *Dendrobium sinense*, a pretty white-and-red flowered orchid that grows on the Chinese island of Hainan. The mystery was solved after Jennifer Brodmann, a scientist at the University of Ulm,[*] mounted a 121-hour vigil to spot potential pollinators. The flowers are actually rewardless, meaning that they don't produce any nectar, but they still received passing interest from 35 insect visitors, 30 of which were a species of hornet that is known to prey on honey bees.

Intriguingly, the arriving hornets didn't just inspect the flower and then buzz off, but pounced aggressively, aiming for the flower's red centre, then abruptly left. A closer look revealed that, in doing so, they were carrying off pollen sacs from the flower and depositing pollen from other members of the same orchid species. But why would they do this when they stood to gain nothing in return?

The answer, it turns out, is a story of sexual subversion never seen before in nature. To get to the bottom of it, the team used solvents to

[*]   *Current Biology* 6 August 2009: Vol. 19. no. 16, pp. 1368–72 DOI: 10.1016/j.cub.2009.06.067

extract and analyse the odorant chemicals from the flowers. They then presented each of the chemicals they had isolated to the antennae of a hornet which had been wired up to an 'electro-antennogram'. This meant that when a chemical was present to which the antenna was sensitive, it produced electrical discharges which the team could measure.

Five chemicals were found in the orchid scent that produced a positive response, which were benzyl alcohol, benzyl acetate, octadecan-1-ol, eicosan-1-ol and 11-eicosen-1-ol. This latter molecule made the researchers sit up like they'd been stung – because it's used by bees as an alarm pheromone and is highly attractive to foraging workers.

Hornets, it seems, have learned to home in on this smell in order to capture bees, which they can then feed to their young. But occasionally, fooled by the irresistible odour of a ready meal, they grab an orchid instead, pollinating it in the process!

A hard pill to swallow: why vitamins might be the death of you

The vitamins, minerals and health-supplements market is now a multibillion-dollar international industry. What we're all buying into every time we take a packet of horse-sized pills off the shelf is the idea that the contents will make us healthier, less likely to catch something and hopefully live longer. It's a bit like a cross between a chemical comfort blanket and an insurance policy – the contents absolve the user of having to worry about doing other boring 'good for you' things, like eating properly, taking regular exercise or not drinking too much.

But the health hype hides a metabolic myth on a massive scale: vitamin tablets don't make you live any longer and they might even shorten your life. At least that's the conclusion of Goran Bjelakovic and his colleagues at the University of Copenhagen,[*] where they've carried out the

* *Cochrane Database of Systematic Reviews* 2008: Vol. 2.
  DOI: 10.1002/14651858.CD007176

largest analysis yet on the pill-popping effects of vitamin supplements.

What the Copenhagen team have done is a meta-analysis. This is a process by which researchers combine the results of many (in this case 67) smaller trials to achieve what is effectively one very large dataset of participants. Doing this makes a study much more powerful, so researchers can iron out bias and spot more subtle trends that might previously have been masked by statistical 'noise' in a smaller study. When Bjelakovic did this, the summed results from over 230,000 people, which included both healthy individuals and people with a variety of ailments, showed that taking antioxidant vitamins such as vitamin A, beta-carotene, vitamin E, vitamin C and selenium, was no better at cutting mortality rates than taking a placebo. The vitamins were, however, much better at emptying patients' wallets.

The bad news for the 'nutraceuticals' industry didn't stop there. When the team focused on 46 trials that were judged to be the most reliable, they found evidence that some of these vitamins may actually be significantly increasing mortality rates. Vitamin A, for instance, was linked to a 16% increase in deaths, and vitamin E to a

Stripping Down Science

4% increase. However, the use of antioxidant vitamins in staving off disease and the effects of ageing is actually based on sound science. Reactive products of metabolism, such as free radicals, are known to damage cells and DNA, and antioxidants can act as a shield, soaking up the radicals before they can pounce destructively on a gene or a cell membrane. But taking excess quantities of these chemicals clearly doesn't work, presumably because there are other knock-on metabolic effects that undo any benefits. In some respects it's a bit like over-filling your car engine with oil and expecting it to run better.

There's no quick fix. The bottom line is that the human body has evolved to absorb and use these micronutrients in the context and concentrations in which they are found in a normal diet, not a packet of pills. The best solution is to do what the nutritionists have been telling us for decades and what public health doctors have recently confirmed in large-scale trials: take regular exercise, drink in moderation, don't smoke and above all, eat five portions per day of fruit and vegetables.

Makes me feel ill just thinking about it . . . maybe I need a supplement.

FACT BOX

Ways to live longer that *are* evidence-based

For those seeking to live longer without risking life, limb and bank balance into the bargain, scientists at Cambridge University have identified four simple things that could gain you an extra 14 years. Between 1994 and 1997, epidemiologist Professor Kay-Tee Khaw and her colleagues[*] recruited 20,000 men and women aged between 45 and 71 years into a study to look at the impact of various lifestyle factors on longevity and health.

At the time of enrolment, the study subjects filled in a simple questionnaire about their lifestyles and were also assessed by nurses at a clinic. They earned a point for every positive answer to being a non-smoker, a light drinker and regular exerciser. They also earned a point if a blood test revealed a vitamin C level consistent with eating about five daily

[*] *PLoS Medicine* 8 January 2008: Vol. 5. no. 1, p. 12
DOI: 10.1371/journal.pmed.0050012

portions of fruit and vegetables.

The team then followed up the volunteers until 2006, tallying up who lived and died. The results showed that, over an average period of 11 years, people who had scored zero points at the initial assessment were four times more likely to have died than participants who had scored four points. They also found that people scoring zero had about the same risk of dying at any time as someone 14 years older than them who had a score of four points.

The moral of this story is quit smoking, drink in moderation, take the stairs not the lift and follow the five-a-day plan (that's portions of fruit and vegetables, not pints or hamburgers) and you might live 14 years longer!

Clone Alone: why modern technology won't bring back your pet pooch

A little while back, a South Korean company, RNL Bio, opened its doors offering a commercial canine-cloning service.* Their first order, with a price tag of $150,000, was to re-create 'Booger', who had died 18 months earlier. He was a pit bull terrier previously owned by a Californian woman who presumably missed him so much she wanted him back again.

You can understand her sentiments. Most of us have mourned the passing of a pet at some time in our lives, but now some seem to be prepared to go as far as cloning their dead dogs, despite the enormous cost, in the belief that they can turn them into the canine equivalent of Lazarus. Unfortunately, the misconception under which most of these people are labouring is the belief that not only will they get back a pet that looks like the one they lost, but it will also have the same lovable

*   rnl.co.kr/eng/main.asp

personality traits, temperament and training.

This is just not true. It's like assuming that a newborn baby would know how to talk because its mother and father could. Similarly, you wouldn't expect a newborn dog to know how to herd sheep because its father was a collie! Although some aspects of behaviour are driven by genes, the majority is learned, so it's wrong to assume that if you clone your dead dog you'll get anything other than a totally new pet that merely looks like your old one. The best evidence to prove this point is to look at identical twins, which are natural clones formed when a developing embryo splits in two. Although they're very difficult to tell apart physically, anyone who is a twin or knows a pair very well will tell you that they certainly have their own unique preferences, likes and dislikes.

The reason for this is that humans' and animals' brains are moulded and shaped by their day-to-day experiences. As the Oxford University neuroscientist and author Susan Greenfield put it to me, 'the brain you go to bed with is not the same brain you woke up with this morning'. So if two clones, dog or human, have different experiences in life, they will be two totally different individuals.

And since it would be almost impossible to ensure that a cloned pet is raised in an identical environment and have the same experiences as its predecessor, it's almost certain that the two animals would be different. The nail was driven further into this coffin recently when scientists showed that random chance also plays a major role in the way some parts of the nervous system wire themselves together. Even if the environment and upbringing were the same, the likelihood is that the resulting brain wouldn't be.

But for pet-parents who are so far undecided about whether to bring their dog back from the dead, there is another option. US company ViaGen run a 'preserving your pets' service.* For

---

*   www.viagen.com/en/our-services/preserving-your-pets/

$1500 and a $150 annual retention fee, they'll store your pets' genes for you so you can clone them later if you like. Even if they choose not to, owners can sleep easy in the knowledge that they've kept their pets' genetic legacy alive, even after the pooch has passed on.

FACT BOX

Genes that control canine coiffures

Dog owners unsatisfied with the concept of merely cloning a canine, preferring instead the prospect of some additional genetic manipulation, will be heartened to know that scientists have recently tracked down the genes that give dogs their hairstyles.

The discovery was made by National Institutes of Health (NIH) researcher Edouard Cadieu and his colleagues in the United States,[*] who successfully sniffed out three crucial genes that control whether a dog has hair that is long or short, wiry or curly, or

[*] *Science* 2 October 2009: Vol. 326. no. 5949, pp. 150–53 DOI: 10.1126/science.1177808

comes with 'furnishings' – the moustache and eyebrows seen in some breeds.

Cadieu and his colleagues made the discovery using a technique called genome-wide association. By comparing the pattern of genetic markers called SNPs (single nucleotide polymorphisms) from many dogs with a single characteristic of interest alongside other dog breeds lacking that trait, the team were able to home in on the genetic region and then the genes underlying the different hairstyles.

Samples from more than 1000 dogs from 80 different breeds led the researchers initially to a gene called RSPO2 (R-spondin-2), which gives certain breeds of dogs their eyebrows. Next, they flushed out a gene called FGF5, which comes in two forms, one of which (the T form) leads to long hair when two copies of it are present, and a G form that triggers a short back and sides. Finally, the scientists were able to track down the gene for curly hair, KRT71, which encodes the protein keratin, used to make the hair itself (and claws). This also comes in two forms, T and C, with curly coated pooches carrying two copies of the T form. Overall, the scientists were surprised to see that the three genes they'd identified accounted for more than 95% of the hairstyles seen in all of the animals analysed.

Also quite surprising was the finding that when the researchers studied the genetic sequences of three grey wolves, none of the alternative hairstyle genes seen amongst the dogs were present in these near relatives. The wolf genes were for short, straight hair and no 'furnishings', suggesting they are a 'wild-type' genetic stock from which modern dogs derive originally.

The difference between a dog and a wolf amounts to about 15,000 years of selective breeding by humans. The fact that just three genes account for the coat differences across such a broad range of dog breeds shows that, in each case, single mutations have produced each coat trait and have subsequently been bred by us into all of the other dogs around today.

FACT BOX

Cats have their owners under their paws

Those who prefer to keep a cat rather than man's more traditional best friend, beware. Your moggy may well be resorting to mewing-related manipulation to tap into your sensitive side.

According to Sussex University scientist (and cat-owner) Karen McComb,[*] there are two 'flavours' of purring: one when a cat is contented and a second 'solicitation' sound produced with 'purr-pose' such as when the animal is after something. 'In my cat's case, it's usually at 5 am when it wakes me up wanting to go out,' says McComb.

To find out whether humans fall for this feline trick, she recorded the purrs of 10 cats and played them to a group of 50 volunteers, which included both cat-owners and non-cat-owners. Participants from both groups pricked up their ears at the solicitation purrs

[*]   *Current Biology* 14 July 2009: Vol. 19. no. 13, R507–08
       DOI: 10.1016/j.cub.2009.05.033

and judged them to be more urgent-sounding, distracting and unpleasant.

To find out why, McComb acoustically analysed the two purr types and found a significant difference. A contented cat produces a steady purr that rumbles along at about 27 Hz, just above the threshold of human low-frequency hearing. But a cat craving attention or food adds an extra sound – effectively a 'mini-meow' nestled inside the ongoing purr. Intriguingly, this 380 Hz additional sound is within the same frequency range as a human baby's cry, which we all know from bitter experience is very distracting. The effect also appears to be more manifest amongst cats that have a close one-to-one relationship with their owners.

This suggests that, once they get to know you, cats craftily tap into a sonic sensitivity linked to your child-nurturing instincts, meaning you're more likely to feed them than throw them out . . .

Boom times are bad for you

Recession, depression, repossession, febrile economy, credit crunch, quantitative easing . . . Few have escaped the headlines, or the impact, of the financial crisis that's gripped the world in the latter part of the 'noughties'. Given the grim state of the world economy, you'd be forgiven for thinking that just about the only people who might profit from a recession are bailiffs, bankruptcy lawyers and newspaper publishers with a penchant for monetary puns and wordplays. But there is a silver lining to the present gloomy outlook: your health is likely to be a beneficiary too!

This sounds counterintuitive, because we're always being told that prosperity makes people live longer. But it turns out this is a myth of almost similar proportions to Barack Obama's US rescue package. Instead, it's official: new work has shown that, alongside share prices, death rates drop during a recession.

This relationship was revealed when Jose Tapia Granados and Ana Diez Roux, both based at the

University of Michigan,[*] compared mortality rates with the state of the US economy over the period straddling the 1930s Great Depression, with striking results. In the years 1923, 1926, 1929 and 1936–37, they report, there were economic booms. But in each of these years, the population mortality rates also peaked. However, in 1921, 1930–33 and 1938, which were all recessions or depressions, the death rates among children and adults fell to their lowest levels. Initially this seems surprising because one would assume that, during recessions, people would be financially stretched, have little money for healthcare and healthy living and would be generally more stressed, all of which would add up to a higher risk of dying. It seems the reverse is true.

Some have suggested that this charted relationship reflects nothing more than a 'lag' effect, whereby people become ill during a recession but by the time they die, the economy is booming again. The researchers discount this theory on the grounds that the timing just doesn't work, because the periods between boom and bust are not constant each time yet the mortality

*   *PNAS* 13 October 2009: Vol. 106. no. 41, pp. 17290–95 DOI: 10.1073/pnas.0904491106

rates change directly in step with the state of the economy.

They point out that, although a fast-growing economy might be good for your bank balance, it's potentially very bad for the health of the person in the street because people tend to work longer hours during a boom and have more disposable income to spend on a deleterious lifestyle, including alcohol and cigarettes, the consumption of which is known to increase at times of prosperity. There is also more traffic, more pollution, and industrial accidents are more common. People also migrate during booms in pursuit of lucrative high-paid work, which can lead to social isolation, itself a risk factor for poor health.

Together, these factors add up to more ill-health during the so-called good times. So rather than rue the recession, welcome it with open arms . . . and see how long your bank manager will swallow the story that your empty account and maxed-out credit card are all part of a healthier lifestyle!

FACT BOX

Pointing the finger at the top market mover to get the best from boom and bust

Love it or hate it, boom and bust is part of a capitalist economy. But to whom should you entrust your money to ensure the best return? The answer would appear to be, when assessing your brokers' credentials, take a tape measure, because scientists have revealed that the relative lengths of a person's fingers can predict money-making ability in some financial markets.

Cambridge University Judge Business School researcher John Coates, who also made headlines by showing that testosterone levels amongst city traders were linked to their daily profits, took photocopies of the palm-prints of 44 financiers.[*] He measured the lengths of the index and ring fingers to find a strong association between finger length and profit

*   *PNAS* 13 January 2009: Vol. 106, no. 2, pp. 623–28
     DOI: 10.1073/pnas.0810907106

or loss. Traders with an index finger shorter than the ring finger were more successful, on average, and the larger the ring finger relative to the index finger – known as the 2D:4D ratio – the more money they made.

'This is an index of testosterone exposure during development,' points out Coates. 'Some of the same genes that control limb and hand development in the embryo are also involved in the development of the urogenital system, so finger length is an index of pre-natal testosterone levels. Testosterone therefore seems to pattern the body and behaviour later in life. We see the same relationship amongst sportsmen playing testosterone-charged sports.'

In the present study, the researchers looked at a specific group of traders who aim to profit by gambling on second-by-second and minute-by-minute fluctuations in the values of certain assets. 'These high-frequency trades require the same testosterone-fuelled rapid reactions that benefit an athlete on the sports field,' says Coates, 'although not all financial roles benefit from high testosterone: some money-making schemes require traders to take a much longer-term view, and high testosterone is unlikely to be of benefit under those circumstances.'

The bottom line is, when making short-term bets on stocks and shares, measure your brokers' fingers . . . ideally before you threaten to break them!

Food for thought: diet-sized snacks make you eat more

A common belief is that if you're trying to lose weight or give up smoking, then buying in bulk is a bad idea. Psychologists say that big bags of food encourage you to eat more, so smaller portions should be the order of the day. However, recent research carried out on snacking Dutch students suggests that the reverse might be true.

Working on the premise that consumers might be lulled into a false sense of dietary security by small portion sizes, University of Tilberg scientist Rik Pieters and his colleagues* set out to discover how a diet-conscious mindset and a big bag of crisps can influence a person's eating habits. The team recruited 140 male and female students who were told they were taking part in an advertising study, which involved watching some television commercials. The participants were given either two large bags or nine small bags of chips to

*   *Journal of Consumer Research* October 2008: Vol. 35. no. 3
    DOI: 10.1086/589564

munch during the viewing. Before the screenings started, half the students were also made 'diet aware' by weighing them in front of a mirror.

The team then totalled up how many bags of chips the students opened and what weight of food they ultimately ate. The results were quite literally gob-smacking. Amongst the non-diet-conscious volunteers, 50% opened the big bags whilst 75% opened the small bags. Weighing the food eaten, however, showed that both groups had nonetheless consumed about the same amounts.

But amongst the participants who were weighed first, the story took an interesting twist. In this case only 25% of the students given big bags of crisps actually opened them. But when they did, they ate *half* as much as the 59% of students who opened their small bags. This shows that, far from helping people to exercise dietary restraint, under certain circumstances restricted-size packages can actually trigger *increased* eating. The team think that the effect occurs because small portions effectively fly beneath the nutritional radar and fail to trigger our normal self-restraining behaviour.

'The tendency of consumers to believe that smaller quantities of tempting products are "acceptable" and to consider single-serving packages even as helpful self-regulatory tools can contribute to increased consumption compared to when products are offered in quantities considered to be "unacceptable", which could instigate consumption restraint,' say the researchers. So big bags in the hands of diet-conscious individuals made them think twice about opening them in the first place, and then made them regret every morsel. The smaller bags, however, didn't trigger the same calorie-sensitive alarm bells and so the students ate more.

According to Rik Pieters, the apparent willingness of food manufacturers to provide 'healthier' and 'smaller' portions may be because they already know this happens, and that by marketing their products in this way they can effectively kill two birds with one stone. On the one hand they can expand their markets by selling more food and on the other they can promote themselves in a healthy light. Consumers, meanwhile, merely expand their waistlines . . .

How the cookie crumbles: you can stay slim by thinking about lunch

Researchers have found that, contrary to expectations, thinking about food can help to reduce how much you eat. We usually associate food fantasies with hunger pangs and a subsequent binge, but a recent study by Birmingham University researcher Suzanne Higgs[*] suggests that this is a nutritional myth. She and her colleagues invited a group of 47 healthy (non-obese) female students to take part in a 'snack-attack' test advertised on the university campus as an investigation into the relationship between food and mood. In reality, it was the culinary equivalent of an episode of candid camera, during which the researchers scrutinised the eating habits of the participants to see whether thinking about food actually made them hungrier.

The volunteers were presented with plates of

[*] *Physiology & Behavior* 9 June 2008: Vol. 94, no. 3, pp. 454–62
DOI: 10.1016/j.physbeh.2008.02.011

biscuits to eat, either one hour or three hours after they were given a set lunch at the laboratory. Before embarking on the biscuit sampling, half of the students were asked to produce a detailed description of the lunch they had eaten, while the other half were asked to write down the details of their journey in to the test that day. After some cookie-related questions intended to mask the true nature of the trial, the participants were invited to eat as many of the remaining biscuits as they wanted. The team then weighed the plates to find out how much each of their subjects had consumed.

Although there were no differences in how hungry they claimed to be, the students who recalled lunch, Higgs found, ate about a third fewer biscuits than those asked to recollect their journey to the lab. The effect was also much more pronounced when the students were tested three hours after lunch compared to when the test was carried out one hour post-lunch.

The researchers think that this is probably because the memory of eating lunch was still vivid enough at the one-hour stage to affect the women's appetites, regardless of what they were asked to recall. By three hours, though, those memories had faded, except in the participants who reminded themselves of what they'd eaten. For them, recalling their last meal did make a difference to how hungry they felt.

So why do we normally associate thinking about food with feeling hungry? Because, say the researchers, this involves thought about food in general. The difference in this study was that the subjects were thinking about a specific meal and how this relates in time to their present situation. Replaying that memory has the power to alter how the brain weighs up appetite, translating into reduced food intake, even if we don't feel less hungry.

Now the researchers want to know whether the process also works the other way around. 'I'm following up with a study that examines the theory in reverse, so whether disrupting people's memories while eating – by watching television whilst eating, for example – causes an increase in appetite,' says Higgs. She didn't, however, mention the possible effects of watching a television show about cooking . . .

There was no greenhouse effect!

A big scientific mystery is how the early earth, four and a half billion years ago, managed to remain so warm and was an ideal place to spawn life, despite the young sun pumping out 30% less light and heat than it does today. Moreover, as the sun has warmed up since, why has the world remained at a steady temperature?

Scientists thought that the answer to this 'faint young sun' paradox was that the planet was initially blanketed in a thick shroud of CO_2 that, through a greenhouse effect, kept the planet artificially warm. But new research suggests that this CO_2 story is a myth, if not a load of hot air, and that the real answer is much more cunning. In questioning, 40 years ago, how the world kept itself warm when it first formed, the American space scientist Carl Sagan effectively threw a planet-sized spanner into the cosmological works. He pointed out that when the solar system first assembled itself, the low level of output from the immature sun would have left earth in the deep freeze and certainly far too cold for anything other than ice to exist.

Yet the geological record paints a very different picture. Written into ancient rocks are clear signs that the earth was bathed in large amounts of water and experienced stable temperatures of 70 degrees Celsius or so. Because there wasn't much oxygen around during this time, scientists initially suggested that a dense smog of methane and ammonia was responsible for trapping more heat from the sun and keeping the temperatures up. But when they realised that ammonia breaks down in sunlight, they gave up on that answer.

The problem was finally thought to have been solved when the American scientist James Kasting[*] suggested in the early 1990s that CO_2 might be the answer. By acting as a greenhouse gas, and with an atmospheric concentration approaching 30% (70 times today's levels), this, together with water vapour, could keep the earth's temperature on track. Kasting provoked a big scientific sigh of relief and, for a while, everyone was happy with his explanation. But then geochemists unearthed ancient rocks that seemed to show that the CO_2 levels were actually far lower than would be

[*]   *Science* 12 February 1993: Vol. 259, no. 5097, pp. 920–26
      DOI: 10.1126/science.11536547

required and the jury was out again.

Now University of Copenhagen scientist Minik Rosing and his colleagues[*] think they have solved the mystery once and for all. They've analysed compounds of iron found in rocks more than 3.8 billion years old. These so-called banded iron formations contain two different iron minerals, magnetite and siderite, which form in different ratios according to how much CO_2 is around. The results clearly indicate that there couldn't have been much more than about three times the present-day levels of CO_2 in the ancient atmosphere, which is a far cry from the 30% that would be needed if CO_2 was the answer.

Instead, their calculations suggest that the effect is down to albedo, which is the amount of solar energy reflected off the planet's surface and back into space. On the early earth, the continents were much smaller, most of the planet's surface was occupied by heat-hungry water, and there were fewer reflective clouds in the sky owing to a lack of sulphur and other compounds that normally trigger cloud formation. Together, these effects meant that, during these early times,

*   *Nature* 1 April 2010: Vol. 464, pp. 744–47 DOI: 10.1038/nature08955

Stripping Down Science

far less of the sun's heat was bounced back into space, so the planet simmered beautifully.

But why didn't the world continue to warm as the sun matured and began pumping out more heat? Because, the researchers show, as this was happening, the continents grew, reflecting light back into space. Life appeared, modifying the atmosphere, and more light-reflective clouds formed in the sky. This, they say, all adds up to the balmy, stable temperatures we enjoy today.

Apart from solving a longstanding mystery, these results are also important, say the scientists, because they show that, contrary to popular belief that CO_2 levels in the past have been much lower than they are today, atmospheric CO_2 concentration appears to have remained relatively stable throughout the lifetime of the earth.

So, when designing computer models to predict potential future climate changes based on past measures of CO_2, scientists need to take this into account. Otherwise, to paraphrase the famous Spanish philosopher George Santayana, if we don't learn from history, we're doomed to repeat it.

FACT BOX

Why we can't rely on trees to combat climate change

We tend to regard trees as allies in the fight against global warming because they lock away carbon dioxide from the atmosphere and encourage the formation of clouds, which help to reflect heat back into space and keep the earth cool. This is part of the argument put forward by Minik Rosing to solve Carl Sagan's paradox (see 'There was no greenhouse effect!').

Unfortunately, scientists have now discovered that when the temperature rises, plants up their output of a class of chemicals that cut cloud formation, potentially accelerating, rather than mitigating, global warming. Normally, microscopic airborne particles called CCNs – cloud condensation nuclei – encourage airborne water vapour to condense into droplets by providing a surface upon which they can form. A major source of these CCNs are volatile chemicals called monoterpenes, which are released in large amounts by coniferous trees.

Airborne monoterpenes react with ozone and also with hydroxyl (OH) radicals, both of which are naturally found in the air at low levels, to produce the chemicals that ultimately link up to form CCNs. Predictably, the greater the concentration of CCNs, the greater the number of water droplets that can form and the smaller they tend to be, which means more clouds to reflect heat and light back into space, helping to offset global warming. And since plants produce more monoterpenes in the summer time, scientists had assumed that, if global temperatures rise, trees would stay in step and increase their output too. This, in turn, would lead to more clouds and more cooling, helping to reset the balance.

But now Astrid Kiendler-Scharr, a researcher based at the Jülich Research Centre in Germany,[*] has found a serious fly in the atmospheric ointment. Using a special sealed chamber to monitor the chemicals

[*]   *Nature* 17 September 2009: Vol. 461, pp. 381–84
     DOI: 10.1038/nature08292

being produced and consumed by plants as they grow, she has found that deciduous trees also pump out a small hydrocarbon molecule called isoprene, chemical formula C_5H_8. And the warmer it is, the more isoprene the trees make. The problem is that this also reacts with the hydroxyl (OH) radicals in the air, so it competes with the monoterpenes that are trying to form CCNs, meaning fewer clouds and therefore less heat reflection back into space, which could accelerate global warming.

Scientists are also worried because, as well as boosting isoprene levels, another study has found that higher temperatures could also transform the Amazon rainforest from a prodigious carbon consumer into a massive carbon dioxide source. Leeds University ecologist Professor Oliver Phillips, together with an international team of more than 60 collaborators,[*] looked at 136 plots of rainforest over a number of years to work out how much biomass (biological matter) was present and

*   *Science* 6 March 2009: Vol. 323, no. 5919, pp. 1344–1347 DOI: 10.1126/science.1164033

therefore how much carbon the Amazon was actually locking away every year. They found that, in the years leading up to 2005, the Amazon was a powerful carbon 'sink', tying up more than a tonne of carbon per hectare of rainforest per year.

But in 2005 there was a severe drought that led to the death of many trees and growth arrest among the survivors. This meant that, as the dead material began to break down, and without fresh carbon being locked away by growth, patches of the forest began losing up to two tonnes of carbon per hectare per year. The significance of this result is that the 2005 drought was provoked by warmer-than-normal water in the north Atlantic, which is what triggered Hurricane Katrina and led to the flooding of parts of New Orleans. Unfortunately, it had the reverse effect over the Amazon, and if global warming continues we might therefore see a drier Amazon more often.

This would mean that the billions of tonnes of carbon locked away by the rainforest every year would cease to be removed from the atmosphere, while at the same time the Amazon would become a net carbon source as the existing biomass broke down. The result would be a dramatic acceleration of the greenhouse effect, with predictable global consequences.

'This should provoke a re-think of the political agenda,' says Phillips.

Coats of many colours

Chameleons have earned a reputation for being masters of disguise because of their incredible ability to change body colour within seconds. But it's a myth that they do so in order to blend in with their surroundings. In fact, the main reason chameleons change colour is so that they can communicate with each other and regulate body temperature. A calm chameleon, for example, is usually pale green, while an angry one turns bright yellow. A cold creature often takes on a darker hue to soak up more sun, while a speed dating chameleon, with mating on its mind, usually sports an explosive burst of reds, greens, browns, whites and blues. But how do they perform the lizard equivalent of a total body makeover in just a few seconds?

The answer, it turns out, is only skin deep and is made possible by a series of specialised pigment-loaded cells, called chromatophores, which are wired to the animal's brain and are also sensitive to hormones circulating in the bloodstream. These chromatophores behave like the pixels of a TV screen. They are arranged in layers beneath the chameleon's transparent outer skin. Those in the uppermost layers are referred to as xanthophores and erythrophores, and contain respectively yellow and red pigments including carotenoids, which are the molecules that make carrots look orange.

Deeper down in the skin is a layer of iridophores. These contain a colourless crystalline compound called guanine that reflects blue light. Then, deeper still, is a layer of cells called melanophores. These contain melanin, the same substance that gives humans a suntan. By varying how active they are, the chameleon can trigger the melanophores to soak up more or less light to make itself look lighter or darker. The pigments within the chromatophores are stockpiled as a collection of granules in the cells. When the chameleon wants to change colour, signals from the nervous system, together with chemicals in the blood, activate individual chromatophores and cause the granules to disperse across the cell and change its colour, rather like giving it a coat of paint.

By activating different combinations of chromatophores at the same time, different colours can be produced. This works in the same way that mixing red and yellow produces orange. So to make itself appear green, a chameleon switches on the yellow in its xanthophores and the blue in its iridophores: blue plus yellow makes green. Or, to make itself appear darker, the animal can activate its melanophores, which allow melanin to spread across the cell and soak up light, turning the chameleon brown. While chameleons may not be masters of disguise, they are certainly makeover masters, at least when it comes to skin redecoration.

Deer-dating data reveals 'survival of the fittest' is a myth

In 2009, the world celebrated the 150th anniversary of one of the most famous books ever published, Charles Darwin's *On the Origin of Species*. Darwin's insight, stimulated by the five years he spent as the naturalist aboard HMS *Beagle*, was to recognise the workings of the process of natural selection, the cornerstone upon which the theory of evolution is founded.

During natural selection the favourable characteristics of an organism, such as those which enable it to succeed in a particular environment, become more common in successive generations. This is because a successful organism is much more likely to breed and multiply than a less successful one, so any unfavourable traits are progressively weeded out, while good traits are concentrated within the population. This is the origin of the term 'survival of the fittest', which was coined in the 1860s by the British economist Herbert Spencer after he read Darwin's book.

Now, though, a new study on wild Scottish deer carried out by Edinburgh University researcher Loeske Kruuk and her colleagues* suggests that the concept of 'survival of the fittest' is something of a myth, or an oversimplification at best. Why? Because scientists had previously overlooked the fact that amongst most animals there are two sexes, and what works well genetically for one sex won't necessarily wash with the other. The researchers made the discovery after several years spent studying the red deer population on the Isle of Rum. During their analysis, the team tested the paternity of the deer using genetic fingerprinting techniques to work out who was related to whom, and then logged how many offspring each animal produced and how many young were then born to those offspring.

What the researchers had expected to see, based on the concept of 'survival of the fittest', was that the successful males should have successful daughters as well as successful sons. But when the results from both generations were compared, astonishingly, a very different pattern emerged. Far from being a genetic paragon of

* *Nature* 28 June 2007: Vol. 447, pp. 1107–10 DOI: 10.1038/nature05912

fitness, the daughters of the successful males who had fathered large numbers of young (both male and female), tended to be relatively unsuccessful themselves in the deer-dating stakes. On average, these females produced fewer calves overall. The sons of the successful males, on the other hoof, were unaffected.

This, say the scientists, is because the genes that single out a successful stag include those for large stature, impressive antlers and the ability to roar the loudest, none of which are of much use to a female. 'What we've found is that the genes that make a successful male do not always make a successful female,' says Kruuk. 'So the idea that some genes are better than others is just too simplistic. Instead it depends on the sex of the individual animal carrying the genes.'

Is the same true for successful females? Do they have wet blankets for sons? 'It's slightly more complicated when you look at it that way round,' says Kruuk, 'because females contribute not only their genes to their offspring's later performance, they also provide maternal care. So a good quality female, although she may be passing on genes which are detrimental in a son, will also provide very high-quality maternal care. These two effects seem to balance one another out.'

Does this leave 'survival of the fittest' for dead, or even stuck in a 'rut'? Perhaps not, because the effect seen in the deer has an unexpected spin-off: it contributes to genetic variation, which makes the population fitter as a whole, since a larger gene pool means a better ability to cope with life's challenges further down the track.

Yes, but can you multi-task?

Time to clock off . . .
if you're a reindeer

Scientists believe that the majority of living things on earth, whether mould, mice or men, use a body (circadian) clock to synchronise their activity to the local environment. In humans, that means we go to sleep when it gets dark and wake up when the sun rises. For nocturnal animals the reverse is true, and many seasonally breeding species track the changing relative lengths of the day and night so they know when in the year to mate. But one thing researchers hadn't bargained for is what happens to animals that live where the sun don't shine – at least for long periods – like the Arctic?

Here, for about six months of the year, the world is plunged into near-continual darkness; for the other six months, the sun never sets. Between the two is an 'equinox' period when days are about the right length, but any animal that attempted to set its clock by the sun in this environment would be in for a chaotic lifestyle, even by the standards of a city commuter. But there are animals that flourish in these conditions, reindeer among them, so what's happening to the body clock in these species? Scientists had suggested that animals like reindeer have evolved merely to ignore their clocks (a trait that seems to be shared with teenagers and students), but this turns out to be a myth that's just timed out.

The idea first came to light when researcher Karl-Arne Stokkan[*] began studying wild reindeer living in his native Norway and on Svalbard, a cluster of islands in the Arctic Ocean, further to the north. Together with colleagues, he equipped groups of the animals with movement-sensitive collars called 'Actiwatches', so he could distinguish when they were moving, and hence awake, from when they were stationary and therefore probably asleep. The results showed that, unlike humans (new parents excluded), reindeer don't sleep through a night but instead have irregular cycles of activity and rest, during which they periodically take a nap. This erratic pattern is maintained regardless of whether it's permanently light or permanently dark, which led the researchers to conclude that the reindeer are more slaves to their stomachs than their body clocks.

[*]   *Nature* 22 December 2005: Vol. 438, pp. 1095–96
       DOI: 10.1038/4381095a

According to Stokkan, being ruminants, the animals' activity is driven by their digestive systems, so they feed whenever the weather permits. Also, by dozing only when they feel tired, there's less chance of missing out on a meal, which is a serious consideration when you live in a land where, rather like a school canteen, high-quality calories can be hard to come by.

The big mystery, though, is how the animals are ignoring their body clocks like this. Originally the researchers thought that, by some behavioural trick, they were able to put the clock mechanism into idle when it suited them. But now circadian scientist Andrew Louden from Manchester University[*] thinks that, for this species at least, the hands have fallen off their biochemical clocks altogether. He and his colleagues took blood samples from reindeer as they were being exposed to artificial short cycles of light and darkness, each lasting two and a half hours. What they saw was that every time the lights went out the levels of one clock-linked chemical, called melatonin, shot up. Normally, melatonin won't change like this if the body clock believes that it's the wrong

[*]   *Current Biology* 23 March 2010: Vol. 20, no. 6, R280–82
DOI: 10.1016/j.cub.2010.02.008

Stripping Down Science

time of day. So the fact that the levels altered in this way suggests that, unlike people doing a boring job, reindeer aren't clock watchers.

To prove the point, the Manchester scientists grew reindeer cells in the culture dish and added DNA from a jellyfish to make the cells glow whenever two of the genes known to power the mammalian body clock became active. Compared with mouse cells, where the genes turned on and off sequentially – like clockwork, you could say – the reindeer cells, unlike Rudolph's nose, refused to light up. This suggests that, for these animals, the body clock has evolved not to tick anymore.

So do they live in a world devoid of a concept of time, destined to be eternally immune to jet lag? The answer is, in terms of day-to-day time, probably yes, but scientists are now discovering that these animals have a far more intriguing hold on long-term timekeeping. According to Louden, 'If you asked a reindeer the time of day – assuming it could speak of course – it probably wouldn't have a clue. But if you asked it the date – in other words, the time of year – they'd probably be able to tell you with extremely high accuracy.' And that's because Louden and his colleagues have found that reindeer seem to have substituted a yearly 'circannual' clock for a daily one. They don't understand in detail how it works, but it seems to involve activating certain patterns of genes in a specialised region of the brain's hypothalamus, which has evolved specifically for this purpose in these animals. This is what enables them to get their migration and mating timetables spot on.

Most importantly, this research answers one of the enduring questions of our time – how reindeer know it's Christmas, and how they cope with staying up all night to keep Santa on schedule . . .

FACT BOX

Body clock

In mammals (other than reindeer) the body clock is based in a specialist region of the brain's hypothalamus called the suprachiasmatic nucleus. This is a small rice-grain-sized cluster of about 20,000 interconnected nerve cells resembling a miniature pine cone. It sits directly above the optic chiasm, the route through which visual information from the eyes is relayed to the brain. A nerve branch from the chiasm feeds information about light exposure from the retina into the clock cells to keep them ticking to time and to reset the system when the clock gets confused by jet lag.

To keep time, the nerve cells that make up the suprachiasmatic nucleus use the molecular equivalent of a genetic domino effect. A linked series of genes turn each other on and off sequentially, taking about 24 hours to complete the cycle, altering the electrical activity of the cells as they go. The nerve cells then transmit this activity to other parts of the nervous system to control how awake we are and the levels of certain hormones circulating in the blood. This, in turn, controls our pattern of sleeping and waking, which is called the circadian rhythm.

Historically, researchers thought that the suprachiasmatic nucleus was the only clock the body had, but more recently that idea has been radically revised. Scientists now know that every cell in the body, with few exceptions, is running the same genetic clock, so every organ and every tissue also knows the time. The difference is that these clocks function as slaves to the master clock in the suprachiasmatic nucleus. They're continuously updated by signals carried by hormones that circulate in the bloodstream, including one called cortisol.

Doctors and scientists now realise how important the body clock is to health and disease. Some drugs, like cancer chemotherapies, work best when given at certain times of the day, stem cell transplants can have different outcomes depending upon when they are given, and mental illnesses, including depression, schizophrenia and some dementias are often associated with poor sleep and tend to improve if the sleep–wake cycle is fixed.

There are also metabolic consequences of disrupting the body's normal timekeeping. People who work night shifts are at higher risk of high blood pressure, strokes and heart disease and, amongst females, the risk of breast cancer increases. Researchers are currently trying to find out why this happens and one way to do this is to put volunteers into sleep experiments. One recent study was carried out by Harvard researcher Frank Scheer and his colleagues.[*]

They placed 10 volunteers in a sleep study environment where they were denied access to watches or clocks and, unknown to the participants, the 'days' were made 28 hours long. This meant that over the 10-day course of the study the participants' body clocks were progressively shifted until they were completely out of phase (equivalent to 12 hours jet lag), and then back into phase with their normal sleep–wake cycles. Throughout the study period, the team collected urine and blood samples and monitored the subjects' blood pressures, metabolic rate and sleep quality to see how this affected the physiology of the participants.

The results were dramatic: levels of the

[*] *PNAS* 17 March 2009: Vol. 106, no. 11, pp. 4453–58
DOI: 10.1073/pnas.0808180106

anti-appetite hormone leptin fell by 17%, with the most pronounced drop at the 12-hour out-of-phase point, the glucose levels of the subjects were 6% higher and their insulin levels 22% higher. The average blood pressure was 3% higher, and the 'sleep efficiency' of the subjects was significantly lower – 67% versus 84% normally. Levels of the stress hormone cortisol, which should peak in the morning and fall during the day, were also reversed, leading to a surge in cortisol at the time when the subjects should have been going to sleep.

This shows that there are genuine and significant metabolic and biochemical effects associated with body clock disruption. And as Frank Scheer and his colleagues point out, discovering what causes these changes will help researchers to come up with effective 'countermeasures' to minimise the health impacts of shift work.

HIV only fatal for humans?

Chimpanzees carry the direct viral ancestor of HIV, the human AIDS (acquired immune deficiency syndrome) virus. This chimp version is called SIV (simian immunodeficiency virus) and is very similar to HIV except that, strangely, scientists have always claimed that, unlike untreated HIV in a human, SIV-infected chimpanzees don't develop an AIDS-like illness and instead, infected animals live alongside the virus without getting sick.

But new research, carried out in the African bush, has found that this claim is actually a microbiological myth and that chimp numbers in the wild – which are already dangerously low – are taking a significant beating at the hands of this virus. Brandon Keele, from the University of Alabama, Birmingham,* made the discovery when he carried out a long-term study on three free-living chimp communities in the Gombe National Park in Tanzania.

* _Nature_ 23 July 2009: Vol. 460, pp. 515–19 DOI: 10.1038/nature08200

Together with colleagues, he collected over 200 urine and 1100 faecal specimens from more than 90 chimps during a nine-year period. These samples were then used to look for antibodies as well as the genetic material of chimpanzee SIV. The team were also able to use genetic fingerprinting technology to identify and follow the progress of individual animals during the full nine years of the study. The results were strikingly at odds with the prevailing view that SIV is a relatively benign infection for chimpanzees.

Infected animals, they found, had between 10 and 16 times the mortality rate compared with uninfected chimps. Females carrying the virus were also less likely to produce offspring and, of those that did, the babies had much higher infant mortality rates compared with uninfected mothers and frequently picked up the infection through their mother's breast milk, mirroring the breastfeeding risk seen amongst HIV-infected humans.

The team also witnessed nine 'new' SIV infections amongst the animals they were following during the study and, by genetically sequencing the viruses involved, found that, just as in humans, most new cases arise through infection passing from individuals that have only recently been infected themselves. This is because when chimps first acquire the virus, they initially develop very high blood levels of the agent, making them much more likely to transmit SIV to other individuals during this early phase of the infection. The same is also true of humans infected with HIV. The team were also able to carry out post-mortems on some of the chimps that died during the study. One animal they examined showed chronic muscle wasting and shrinkage of the liver, multiple abscesses in the abdomen and depletion of the key CD4 immune cells targeted by the virus. Tissue samples obtained from other animals also showed similar parallels with human AIDS, including the loss of white blood cells from the spleen.

So, far from being a walk in the Gombe National Park, SIV is a serious and often fatal infection for Africa's chimpanzees. Apart from overturning an inaccurate dogma, this research also helps to highlight an important future avenue of research: chimps with SIV seem to develop the disease in the same way we do, but other primate species with their own forms of SIV do not. If scientists can track down which viral or host factors are responsible for the progression of the disease in some species but not others, it might be possible to uncover new ways to combat the virus in both humans and our next nearest relative.

FACT BOX

HIV biology

HIV/AIDS is, without doubt, the worst pandemic that humankind has ever faced. Researchers estimate that about 25 million people have already died of the disease and that about 35 million people are currently carrying it. In 2008, there were more than 7000 new HIV infections every day – one every 12 seconds.

Untreated, HIV slowly destroys the immune system by attacking key white blood cells that carry a marker on their surfaces known as CD4. When the supply of these cells is exhausted, which occurs after about 10 years on average, the infected person develops AIDS – and becomes vulnerable to a range of lethal infections caused by a host of different microbes ranging from viruses and bacteria to fungi and amoebae.

Scientists suspect that HIV 'jumped' into humans about 100 years ago, following close contact between a person – perhaps a hunter or a butcher preparing bush meat – and the blood of a chimp carrying SIV. Because humans and chimpanzees are so genetically alike, the SIV in the chimp is thought to have managed to infect the exposed human and then mutate within them to become the human AIDS virus. We know that HIV has jumped into humans in this way at least twice, because there are actually two types of HIV in circulation, HIV-1, which came from chimpanzee SIV, and a rarer form, largely confined to West Africa and called HIV-2, which came from a different primate species called a sooty mangabey.

When HIV infects an individual, it first has to break through the body's main defence, the skin. For this reason, sex, needle sharing and the use of contaminated blood products are the main ways in which the virus is passed on.

In the case of sexual transmission, researchers have discovered recently that the virus triggers the cells that make up the surface layers of the genital tract, and other mucous membranes, to part company with each other, creating HIV-sized chinks in the body's immunological armour. This allows the virus to penetrate and infect the susceptible immune cells that sit in the tissue below. Once inside these cells, the virus inserts a copy of its genetic material into the DNA of the cell. This is then used as a template to produce thousands of new virus particles which bud off to infect other cells around the body.

Part of the reason why HIV is so difficult to treat is that the virus uses as its genetic material a chemical relative of DNA called RNA. But unlike DNA, which contains two strands of information, one the mirror image – and effectively a back up – of the other, RNA contains just one single strand. With nothing to check the message against, the virus makes multiple genetic spelling mistakes when it copies itself, introducing mutations that can alter the appearance, virulence and drug-resistance profile of the virus.

To get around this problem, doctors now prescribe cocktails of different drugs which target the virus from several different biochemical directions at the same time. This approach, known as HAART (highly active anti-retroviral therapy) slows down the rate at which resistant mutants appear, helping patients to remain healthier for longer. But drug therapy is only a temporary solution and, owing to cost, one which is available only to a minority of those who are infected. Instead, what the world sorely needs is an effective AIDS vaccine.

So far this has proved impossible to achieve. One of the reasons for this is that to block HIV, the body needs to be able to produce antibodies capable of neutralising the viral velcro employed by HIV to lock onto and invade CD4 immune cells. But the virus keeps this part of its structure concealed beneath a meshwork of sugar molecules and only reveals its hand in the final moments as it infects a cell, making it very difficult for the immune system to intervene.

That said, scientists have found recently that some people can make antibodies capable of blocking this part of the virus, suggesting that if a vaccine can be produced to allow everyone to make these antibodies, then we may finally be looking at a way to stop the 7000 new cases of HIV that are being diagnosed daily.

In the meantime, apart from sending out clear safe-sex messages, a number of other anti-AIDS strategies are being investigated. One study carried out recently in South Africa by Salim Abdool Karim and his colleagues at the University of KwaZulu-Natal, Durban,[*] showed that a gel impregnated with the anti-HIV drug tenofovir and applied vaginally before and after sex reduced the rate of new HIV infections amongst the women using the gel by up to 50%.

Also, amongst men, researchers have shown convincingly that being circumcised can have

*   www.sciencemag.org/cgi/rapidpdf/science.1193748.pdf
     DOI: 10.1126/science.1193748

an equally powerful protective effect; several studies have confirmed that circumcision cuts the HIV infection rate by over 60% in males who undergo the procedure. Removing the foreskin is probably protective because it reduces the mucosal surface area through which infection can occur and also leads to the virus becoming deactivated more quickly because there are fewer warm, wet recesses in which it can lurk.

Less with a bang than a whimper

A long-held view in cosmology circles is that the solar system, comprising the sun, the earth and all of our neighbouring planets, was catapulted into existence about four and a half billion years ago by the arrival of a huge shockwave. A nearby star that had reached the end of its life is thought to have blown itself to pieces in a catastrophic explosion known as a supernova, producing in the process a multimillion mile an hour maelstrom of dust and debris. This cannoned through space until it ran headlong into a cloud of swirling gas.

This cloud would have been about 100 times the present earth–sun distance in diameter and would have weighed about three times as much as the sun. The arriving shockwave is thought to have squeezed the gas together, causing it to begin to collapse under its own gravity, triggering the formation of a central 'proto-star'. This would have been surrounded by a protoplanetary soup of spinning material from which the planets we see today would slowly have emerged as gravity glued the debris together.

A nice theory, but a recent scientific discovery has shown that it's a myth. If this model of earth's birth were true, then remnants from the early solar system, such as ancient meteorites, ought to contain the signature of radioactive iron compounds known as iron-60, which would have been blasted from the innards of the star that exploded. But when Copenhagen University researcher Martin Bizzarro and his colleagues* went looking for this iron-60 (in the form of its radioactive breakdown product, nickel-60), they couldn't find any. In fact, younger meteorites, formed after the solar system had come into existence, contained more iron-60 than the older ones. This totally turned the supernova theory on its head, because the story told by the meteorites suggested that a nearby star had indeed blown up back in history, but it happened after the solar system had already formed.

So what did spark us into existence? Luckily, there was another chemical clue lurking inside the samples that the team analysed. Both the young and old meteorites contained another element, aluminium-26, and this can only mean one thing:

* *Science* 25 May 2007: Vol. 316, no. 5828, pp. 1178–81 DOI: 10.1126/science.1141040

that a super-massive star, at least 30 times the size of our own sun and with a lifetime of just a few million years, must have existed in our cosmic backyard during the time when the solar system was forming.

Stars on this scale burn off their fuel very fast and produce a powerful solar wind laden with material from their surface layers, which includes aluminium-26 but not iron-60. 'This rules out the supernova trigger,' says Bizzarro. Instead, he thinks the wind from this giant star probably buffeted the ball of gas that became us into forming the solar system, adding aluminium-26 to the mixture as it went. A few million years later, the star blew itself to pieces, showering the young solar system with iron-60 from its core, thus explaining why the younger meteorites had the hallmark of iron-60 but the older ones didn't.

It looks as if we may have to rewrite the history of how the solar system came to be and on the basis of these findings, we may well have had gentler origins than space scientists first thought.

Nitrous oxide: no laughing matter

Nitrous oxide (formula N_2O) is a volatile gas discovered by the English clergyman and scientist Joseph Priestley in 1793. (Priestley was certainly a bit of a gas man, because he also discovered oxygen, carbon monoxide, carbon dioxide, ammonia and sulphur dioxide.) Priestley made his nitrous oxide by heating ammonium nitrate in the presence of iron filings, and then passing the nitric oxide (NO) that came off through water: $2NO + H_2O + Fe \rightarrow N_2O + Fe(OH)_2$.

Humphry Davy, from the Pneumatic Institute in Bristol, England, then began to experiment with the physiological properties of the gas, and visitors to the institute were given nitrous oxide to breathe. Their reactions, and his own experiments, led Davy to coin the term 'laughing gas', and he also noticed that the gas had anaesthetic properties. 'As nitrous oxide in its extensive operation appears capable of destroying physical pain, it may probably be used with advantage during surgical operations in which no great effusion of blood takes place.'

However, for 40 years or so after Davy made this observation, most N_2O was used for recreational purposes, including at public shows and carnivals where members of the public would pay a small price to inhale a minute's worth of the gas. It wasn't until the mid-1800s that doctors and dentists began to re-explore its painkilling potential.

As is usually the case with a medical breakthrough, it took an accident to help a local dentist make the intellectual leap that was to catapult N_2O into the domain of medicine. Horace Wells watched as one of the volunteers breathing the gas, a man named Samuel Cooley, staggered into some nearby benches and injured his leg. What intrigued Wells was that Cooley remained unaware of his injury until the effects of the gas wore off. Realising that N_2O might possess painkilling qualities, Wells approached the demonstrator, a medical school dropout called Gardner Quincy Colton, and invited him to participate in an experiment the next day.

Colton agreed and subsequently administered nitrous oxide to Dr Wells while another dentist extracted one of Wells' teeth. Wells experienced no pain during the procedure, and the birth of N_2O as a dental and medical painkiller had arrived. That's the history of the gas. Since that time, it's been embraced as a safe agent that can be used for pain relief (such as during childbirth and dental procedures) and in general anaesthesia. On its own, it's not a sufficiently potent anaesthetic to induce (i.e. cause) anaesthesia, but once a patient is 'under', it's a very good gaseous agent for anaesthetic 'maintenance'.

In this respect, N_2O isn't that unusual, since most volatile gases can behave as anaesthetics with intoxicating effects – they differ only in their potency (i.e. how much of them is needed to have an effect). Volatiles with this property include the butane you squirt into your cigarette lighter and even petrol vapours. In fact, this latter example has been a serious problem in parts of Australia, where members of some communities have been sniffing petrol. This has resulted in BP (British Petroleum) recently producing a blend of unleaded for the Australian market that is less suitable for sniffing.

No one knows exactly how general anaesthetics work, but the fact that they are usually organic, lipid-loving chemicals suggests that they probably alter nerve cell function by dissolving in the oily membrane that surrounds our cells and affecting the behaviour of membrane pores or channels which control the excitability of the cell. Alcohol probably works similarly and there is now evidence that it specifically renders cells more sensitive to one of the brain's inhibitory nerve transmitters called GABA. This means that cells become less responsive in the presence of alcohol, which is why booze is a central nervous system depressant.

As an aside, it's not just animals that can benefit from the effects of N_2O. Cars receive a boost in performance when a burst of 'nitro' is injected into the cylinder during combustion. The heat of the burning fuel causes the nitrous oxide to decompose to nitrogen and oxygen: $2N_2O \rightarrow 2N_2 + O_2$. So two molecules of gas turn into three molecules of gas, which increases the volume of products inside the cylinder, boosting performance. A bit like Viagra really, although that relies initially on the effects of nitric oxide (NO), rather than nitrous!

No experience necessary, no immunosuppression needed

When American surgeon and Nobel laureate Joseph Murray performed the world's first kidney transplant in December 1954, he was successful largely because the recipients, Richard and Ronald Herrick, were identical twins. Because the brothers shared the same DNA code, their individual immune systems could not distinguish the organs of one man from the other and so there was no question of rejection.

Unfortunately, the majority of the 6.8 billion people currently alive on earth aren't lucky enough to have an identical twin from whom to beg or borrow an organ whenever they need one. Even if they did, it's unlikely that organs like the heart, of which most normal people have just one, would be volunteered willingly. As a result, the majority of the transplants carried out today are 'allografts'. That is, they involve organs taken from someone who is genetically different from the recipient – often a healthy individual who has died following an accident.

But therein lies the problem. The immune system marshals a highly trained army of white blood cells and antibodies which are programmed to tell friend, known as 'self', from foe. So whenever foreign materials are introduced to the body, with the clever exception of a developing foetus, the immune system can detect that the chemical markers or antigens displayed on the surfaces of the cells in foreign tissue do not match those in the rest of the body. When this occurs, the presumed impostor is attacked and destroyed. If this involves a donor organ, it's termed rejection, a process first described by the French surgeon and Nobel winner Alexis Carrel at the turn of the last century.

To prevent rejection from occurring, transplant doctors try to genetically match donors and recipients as closely as they can, but there will inevitably be differences between the two and so there is always the prospect of an immune attack. This problem held back the transplant field for many years until the 1970s, when the immune-suppressing drug cyclosporine was discovered. Cyclosporine chemically deafens white blood cells to the sounds of their own inflammatory signals, which prevents the immune system from mounting its normally well-orchestrated attack and thus protects the donor organ from destruction.

Sadly, suppressing the immune system in this way comes at a cost, because patients are less able to fight off infections and they are also more prone to malignancies, since the immune system also has a role in killing off cancer cells. Until recently, doctors thought that these side effects were necessary evils and that immunosuppression was vital for the continued survival of a transplanted organ. Now it looks like this might be at least part-myth.

Indeed, Harvard transplant researcher Megan Sykes and her colleagues[*] have discovered that if a patient undergoing a kidney transplant also simultaneously receives a bone marrow transplant from the same donor, together with a drug to temporarily remove all their white blood cells, they can subsequently stop all immunosuppressive drugs without any signs of rejection. 'If the bone marrow of two different individuals exists in the

---

[*]   *New England Journal of Medicine* 24 January 2008: Vol. 358, pp. 353–61

same patient at the same time, the donor bone marrow can re-educate the immune system so it regards tissue from the donor as self,' says Sykes. 'We've done five patients in a pilot study; four are doing very well. They've been off immunosuppression for several years, five years in one case, and their kidneys are being accepted despite the lack of any immunosuppression.'

Although the researchers aren't exactly sure yet of the mechanics of the process, they suspect that cells from the donor bone marrow trigger a process called 'peripheral tolerance', whereby any of the patient's own cells that would normally attack the donor organ instead switch themselves off.

Will the same trick work for other organ transplants? Tests on animals suggest that it certainly can, but at the moment the technique is only suitable for live donor transplants (where a donor gives one of their two healthy kidneys to help a friend or relative). This is because the patient needs to be pre-treated with drugs over a five- to six-day period so that they can receive the bone marrow from the donor.

Since most transplants involving other organs come from donors who have died, it's currently not possible to keep the organs alive long enough while the patient is prepared. Having said that, a new technique being pioneered by Oxford University researchers Constantin Coussious and Peter Friend can now keep livers alive outside the body using an artificial bloodstream, so it may indeed be only a matter of time . . .

Drowning in quicksand? It's a myth

There was a time when almost every action movie seemed to involve the hero or villain becoming swamped in quicksand, sinking away until only their hat remained on the surface. Even Flash Gordon and vine-swinging apeman Tarzan were victims during their careers. But contrary to what Hollywood would have you believe, it's actually impossible to drown in quicksand, but almost as impossible to escape, as a Dutch scientist found when he produced his own homemade variety in the laboratory.

Daniel Bonn[*] was on holiday in the Iranian province of Qom when he saw a sign saying 'Danger: Quicksand'. Local shepherds had told him that camels and people (usually those who had dared to disagree with the local regime) had periodically disappeared in the area. Realising that science didn't actually have an answer to the

[*]   *Nature* 28 September 2005: Vol. 437, p. 635 DOI: 10.1038/437635a

quicksand conundrum, he took some samples home with him.

Analysis of the composition of the 'quicksand' showed that there are four key ingredients: sand (obviously), water, clay and salt. Together, these materials form a structure resembling a house of cards, with large water-filled gaps between the sand particles, which are loosely glued in place by the clay. As long as it's left alone, the structure remains stable. But as soon as it's disturbed by something stepping on it, the clay changes from a jelly-like consistency to a runny liquid. The effect is the same as stirring a pot of yoghurt. Liquefying the clay makes the quicksand about one million times runnier, and the whole house of cards comes tumbling down with you inside it.

Very quickly, the sand sinks to the bottom and the water floats to the top. This is where the salt comes in. When there's enough salt present, as soon as the clay particles liquefy, electrical charges make them begin to stick together to form bigger particles which also settle in with the sand. The result is a very stodgy layer of sand and clay, twice as dense as the original quicksand and packed tightly around the trapped body parts.

So how do you escape? Well certainly not

the way Hollywood would have you do it – by being pulled out by a horse – because Daniel Bonn's measurements show that the force needed to extract a trapped foot (10,000 newtons) is equivalent to that needed to lift the average family car. You'd probably escape, but minus your legs. The best way out is to try to rebuild the house of cards around the trapped body parts. Making small circles with each part of your body re-introduces water between the sand and clay particles, reducing the density and making it easier for someone to heave you out.

Everyone apart from a Hollywood director can take solace from the most important finding of the research: it's impossible to drown in quicksand – you should only sink halfway. The density of quicksand, at two grams per cubic centimetre, is twice the density of a human (one gram per cubic centimetre). Yet to sink, an object needs to be more dense than the stuff engulfing it, meaning humans, being less dense than quicksand, won't submerge completely and instead become mired at roughly waist height. So stuck you might be, but drowned you wouldn't!

Aspirin' chemist

It's a common myth that the popular painkiller aspirin comes from willow trees. In reality, aspirin is one of the world's earliest 'designer drugs'. While it's based on a chemical found in willow bark – a substance called salicin – the drug itself dates from 1897 and owes its existence to the perseverance of two chemists, Felix Hoffman and his boss Arthur Eichengrun, who were employed by the German pharmaceutical giant Bayer. The story goes that when he made aspirin, Hoffman was searching for a more tolerable painkiller for his father, who was crippled by arthritis. Hoffman senior had found some relief in willow extracts which, even in the late 1800s, had been used for pain and fevers for centuries.

The application of this tree-based salve probably stems from the ancient Greek physician Hippocrates, who had described using powdered willow for labour pains. There are also reports, from the 1760s, of an English priest called Edward Stone, who wrote to the Royal Society in London describing how he had 'accidentally tasted' willow tree bark and subsequently found it to be very effective in the treatment of 'fevers and agues'.

The apparent effectiveness of this remedy had provoked early 19th-century scientists to try to identify the chemical responsible. The aim was to find a way to produce it in greater quantities and more cheaply than their present approach, which was to strip the bark off the nearest willow tree, a system which would have involved an arthritis-inducingly large amount of hacking, since the best processes at the time were yielding only 30 grams (about one packet's worth) of the drug from every 1.5 kilograms of bark.

The chemical breakthrough came in 1838 when Raffaele Piria, working at the Sorbonne in Paris, discovered the chemical structure of willow salicin. It turned out to be a sugar molecule linked to a ring of carbon atoms. With some molecular tweaking, Piria was able to replace the sugar molecule with an acid group. The result was a substance that could readily be made in a test tube but retained the painkilling power of the original willow chemical. It was salicylic acid, the ancestor of aspirin.

But there was a problem, and this is where Hoffman junior comes in. Despite its effectiveness at lowering temperatures and relieving pain, provoking widespread public demand for salicylic acid, there were serious side effects: it caused tinnitus (ringing in the ears) and was extremely corrosive to the stomach. People who used it regularly often developed stomach ulcers, including Felix Hoffman's own father.

To make a more 'gastro-friendly' version of the drug, Hoffman tried a chemical trick that had succeeded in taming some of Bayer's other products. This was called 'acetylation' and involved adding a chain of two carbon atoms to the side of the molecule. He applied this to salicylic acid, producing in the process acetyl salicylic acid (ASA), or aspirin. It was subsequently tested on animals and then patients in a hospital in Halle an der Saale. The results were extremely encouraging, largely because it seemed to be much better tolerated than its salicylic acid parent.

Unfortunately, the company couldn't patent their potential new analgesic blockbuster because a French chemist, Charles Gerhardt, had already made it (in a different way) 50 years previously and then dismissed aspirin as being 'of no significance'. So instead, Bayer took a different tack. They marketed what's now the world's most popular painkiller under a trade name, 'aspirin', some say as a nod to the Bishop of Naples, St Aspirinus, the patron saint of headaches.

For Hoffman, life didn't stand still. Two weeks after making aspirin, he invented heroin, but that's another story . . .

FACT BOX

Aspirin

Aspirin achieves its anti-inflammatory and pain-numbing effects by blocking the production of a family of chemicals called prostaglandins, which were first described in the 1930s by a Swedish physiologist, Ulf von

Euler, as secretions from prostate glands (thus the name prostaglandins).

They have a range of effects in the body, some good and some bad. In the digestive tract, prostaglandins boost the production of a protective layer of mucous to shield the stomach wall against the effects of its own acid, and in the kidney they are used to regulate blood flow and therefore the rate of urine production. Another prostaglandin is also pumped out by the linings of healthy blood vessels to prevent platelets (part of the coagulation system) from sticking to the insides of the vessel walls.

Prostaglandins are also produced at sites of injury or infection, where they have an inflammatory effect. They open up blood vessels, making the area red, hot and swollen; they sensitise pain-detecting nerve fibres to make the area throb; and they also act as a beacon for the immune system, causing white blood cells to flock to the affected area. They are made by an enzyme called cyclooxygenase, known as COX for short, which is switched on by tissue damage. The aspirin molecule, acetyl salicylate, irreversibly locks onto this enzyme, permanently deactivating it. This means the body can only make more prostaglandins by synthesising new COX enzymes, which takes time, explaining why the drug works for a while but then wears off and a repeat dose is needed.

Apart from its role as a painkiller, aspirin has also gained a reputation as a major lifesaver in combating heart disease and strokes. This is because it reduces the ability of blood to clot, specifically by targeting blood platelets, tiny cell fragments whose job it is normally to trigger blood coagulation at sites of vessel injury. However, if an artery becomes partially furred up, platelets can be fooled into initiating clotting inside the diseased blood vessel, leading to a blockage that can cause a heart attack or brain damage according to which vessels are involved.

To do this, platelets need to produce an activating substance called thromboxane A2, which is a chemical relative of the prostaglandins. As aspirin also blocks the production of thromboxane A2, it makes it much harder for platelets to stick to things they shouldn't, thereby 'thinning the blood'. Consequently, millions of people who are at risk of heart attacks or strokes take a low daily dose (75 mg) of aspirin, which has been shown to save lives. But there are millions more people who are not viewed as at risk of a stroke or heart attack, yet also elect to pop an aspirin every day, 'just in case', because they believe that aspirin can prevent vessels from becoming clogged in the first place.

This issue was the subject of a recent large study involving 30,000 initially healthy subjects who were given either daily aspirin or a placebo over an eight-year follow-up period by Edinburgh researcher Gerry Fowkes.[*] He found no difference in subsequent heart attack rates between the people who took aspirin every day and those who didn't. But what he did see were twice as many 'major bleeds' amongst the patients on aspirin compared with those not taking it. So, in the absence of a major heart disease risk, aspirin might do more harm than good.

[*]   *JAMA* 3 March 2010: Vol. 303, no. 9, pp. 841–48

Blood clotting aside, two other important health areas where aspirin can also make an impact are prevention of some cancers and Alzheimer's disease. In the latter case, researchers think that aspirin damps down inflammation in the nervous system, thereby reducing the levels of a protein called beta-amyloid that can otherwise build up in the brain, damage nerve cells and lead to dementia.

The same anti-inflammatory effect is also likely to be behind the observed reduction in breast, bowel and lung cancer risk in people who use aspirin for long periods, and researchers have also recently found that taking the drug can cut the risk of the cancers coming back in people with previously-treated malignancies. Andrew Chan, from Harvard in the US,[*] recruited just under 1300 newly diagnosed bowel cancer patients, 549 of whom took regular aspirin. He compared how long these patients survived against 719 individuals with the same diagnosis who did not take regular aspirin.

*   *JAMA* 12 August 2009: Vol. 302, no. 6, pp. 649–58

The aspirin users, he found, had an overall reduction in mortality rate of 21%, and a 29% drop in their mortality from bowel cancer. The benefit was similar regardless of how advanced an individual's cancer had been when they were first diagnosed or how their disease was managed, suggesting that aspirin might be an excellent adjunct to existing bowel cancer therapies.

Why does it work? Unfortunately, this was an observational study, so the team do not know the precise mechanism by which aspirin achieves this significant life-prolonging effect, but interestingly they did find that the patients who enjoyed the most gains from the therapy had tumours that over-expressed a gene for another form of cyclooxygenase (COX) called COX-2. This enzyme also produces inflammatory prostaglandins, so perhaps by preventing their formation, aspirin reduces within tissues the oxidative stress that can damage DNA and encourage cells to become cancerous.

Life, Jim, but not as we know it . . .

Most people believe that everything alive on earth today gets its energy, one way or another, from the sun. The dogma goes that plants capture sunlight and turn it into chemical energy and then other organisms feed on the plants, and ultimately on each other, funnelling the energy up the food chain until it arrives on our dinner plates. Indeed, even bacteria living in the scalding conditions of mineral-rich hydrothermal vents on the ocean floor still rely on the by-products of other organisms elsewhere on earth for their survival.

But now a trickle of water in a goldmine three kilometres beneath the earth's surface has revealed a population of bugs powered not by the sun but by radioactivity. And the finding therefore fuels speculation that life could eke out a similar existence deep underground on planets elsewhere in the solar system and beyond.

The findings have been made by US researcher Lisa Pratt and her colleagues.[*] The team heard

*   *Science* 20 October 2006: Vol. 314, no. 5798, pp. 479–82
     DOI: 10.1126/science.1127376

about a deep fracture that had been opened up by drilling in a South African goldmine not far from Johannesburg. The team visited the site and collected samples of water that were pouring out of the fracture site. When they analysed the water, they found that it had a very ancient chemical fingerprint. In other words, it had been sealed off from the rest of the world for at least three million years and possibly for as long as 25 million years.

Even more exciting was the discovery that the water also contained the DNA profiles of a vast number of bacterial species. One strain of bacterium stood out because it was so abundant. It was a relative of the bugs known as *Fermicutes*, which are found at deep-sea hydrothermal vents. These bacteria can use hydrogen and sulphur compounds as an energy source and, through their growth, feed and sustain other types of microbes.

But where were these fracture-dwelling bugs getting the hydrogen and sulphur with which to sustain themselves in this underground pocket of water for the last 25 million years? The unlikely answer is natural radiation. In the mine, the surrounding rocks contain uranium.

The radioactivity it produces splits apart water molecules, producing highly reactive chemicals called free radicals. These react with minerals, such as pyrites (fool's gold) in the surrounding rocks, to produce hydrogen and sulphur compounds that the bacteria can use.

So, apart from shattering the myth that all life on earth depends upon the sun, these new findings suggest that life could well exist along the same lines elsewhere in space. And even if life as we know it became extinct on a planet where organisms once flourished, it's possible that somewhere beneath the surface there may well be a colony of thriving bacteria, fuelled by a nuclear reaction . . .

Storing up trouble from radioactive waste

Apart from naturally radioactive rocks, another source of radiation is the stuff we churn out to fuel nuclear power stations, reactors aboard boats and submarines, and even the arms industry. Thankfully, it's a simple task to deal with the radioactive waste that's left over, isn't it? We can just turn it into concrete and bury it, can't we? 'Wrong!' says a leading Cambridge scientist.[*]

While nuclear power, being carbon-neutral, may seem like a safe haven in the present climate-change storm, it brings with it a unique problem that, until now, has gone largely unappreciated. That is, what to do with the leftovers. Of course everyone knows that nuclear power stations produce a significant amount of high-level radioactive waste like uranium and plutonium: Britain has 470,000 cubic metres of the stuff loitering in temporary storage and the US is sitting on 50,000 tonnes of 'hot rods'.

[*] *Nature* 11 January 2007: Vol. 445, pp. 190–93 DOI: 10.1038/nature05425

Until recently, we thought that getting rid of it was quite trivial in the grand scheme of things – you just need to encase it in something hard, bury it, and wait long enough for the radioactivity to die away. Admittedly, in the case of plutonium, that does mean waiting at least a quarter of a million years, but what's that between friends?! But therein lies the problem, because nuclear scientists had planned to mix waste plutonium and uranium with a synthetic mineral and fire it to bake the radioactive equivalent of bone china. The resulting ceramic, containing the radioactive atoms safely sequestered inside the crystal structure, could be buried without any risk of the material escaping.

Indeed, the idea has proved sufficiently popular with the US nuclear industry that they've already sunk US$7 billion into a prospective burial site in the Nevada desert called Yucca Mountain. But if they go ahead and finish the job, they could land themselves with a much bigger hole in their pockets than the one made by the anticipated US$100-billion price tag. This is because it's a myth that these ceramics are stable. In fact, research now suggests that they don't even last 1000 years, let alone the required 250,000.

This worrying result came to light when Cambridge earth scientist Dr Ian Farnan and his team developed a new way to see inside 25-year-old samples of these ceramic crystals that contain radioactive materials. What they saw made alarming viewing, because it showed that the crystals were falling to pieces, and much more quickly than scientists had expected.

When plutonium or uranium atoms undergo radioactive decay they spit out an alpha particle, which consists of two protons and two neutrons stuck together. This travels through the crystal and can pull electrons off some of the nearby atoms, but the damage is relatively minor. Far more serious, and what hadn't previously been appreciated, is what happens to the original uranium or plutonium nucleus. When this ejects the alpha particle, it recoils, like the kick of a gun, which sends it careering into other atoms in the crystal, knocking about 5000 of them off kilter at a time and destroying the integrity of the material.

The net result is that after just a few hundred years in storage, the proposed materials would be riddled with cracks and leakier than the *Titanic*. And after just 1500 years, and a far cry from the required 250,000, the material would have fallen apart completely.

However, every cloud has a silver lining, even if it is slightly radioactive in this case. As Ian Farnan points out, it's better to know now than later, when the stuff might already have ended up in the ground. Thankfully, the technique he's developed can be used to identify more robust mineral recipes that can take a heavier radioactive beating, or even repair themselves so that they can go the distance without breaking down. Let's hope so.

It don't necessarily glow, bro!

Simpsons fans will know only too well the opening sequence of the cartoon in which Homer discovers, during his commute, that he's taken some of his work home with him – in the form of a radioactive fuel rod from the nuclear power plant. Unsurprisingly, the lump of material he subsequently throws out of the car window is glowing a ghostly green colour. But basking in that radioactive light is a luminous myth of atomic-powered proportions, because most radioactive substances don't really glow at all, let alone light up green.

The basis of this belief stems from the late 1800s and early 1900s when the Polish pioneer of radioactivity Marie Curie, working in Paris with her husband Pierre, discovered the element radium, which they named after the Latin word *radius*, meaning a ray. Once they began to isolate the metal in reasonable amounts, the Curies noticed that it appeared to emit an attractive blue glow. According to biographer Barbara Goldsmith, Marie described the ethereal glow as 'fairy-like' and kept a jar of pure radium salts beside her bed, where it presumably functioned as an attractive, albeit potentially lethal, radioactive nightlight!

However, contrary to popular belief, the glow from the tube wasn't the radiation per se, but rather the effect it was having on other chemicals that were also present. This is because as radium decays, it spits out energetic particles and waves including alpha particles, beta particles and gamma rays. When these pass through a material they knock negatively charged electrons off some of the atoms, triggering a process called ionisation. Most of the electrons liberated like this subsequently snap back onto their parent atoms, but first they have to surrender the extra energy they gained to start with. To do this they pump out light, some of which is in the visible part of the spectrum, meaning we can see it.

Chemicals that behave like this, by emitting visible light when they are excited, are referred to as phosphors. As it turns out, Marie Curie's bedside sample would have contained the radium compounds radium bromide, radium fluoride and radium chloride, which are themselves phosphorescent. So when they were hit by the radioactive emissions of the radium itself, they glowed, producing the blue light she found so soothing.

But if Curie's light was blue, where does the idea that radioactive chemicals glow green come from? The most likely source was the 1902 invention, by the American engineer William J. Hammer, of a radiation-powered glow-in-the-dark paint. Researchers found that the substance zinc sulphide would glow bright green when it was excited, including when it was exposed to radioactivity. The glow could also be made brighter if a small amount of copper was added. So Hammer put two and two together and mixed in some glue and a small amount of radium, producing a paint that was kept permanently illuminated. Overnight, quite literally, reliable glow-in-the-dark products were born.

Scientists subsequently tinkered with the chemicals to produce a range of different colours, but the greens were the easiest for the eye to see, so those were the ones that stuck, together with the myth that radiation glows in the dark. Radium-powered watch and instrument dials continued to be produced into the 1960s. Today, safer alternatives to radium have been found, together with better materials that can soak up the energy in natural light and release it very slowly to produce glow-in-the-dark products that shine for much longer without the need to resort to radioactivity. That said, radiation-powered fluorescence still has its place, mainly on the hands of certain expensive watches and in military equipment, although these tend to use the safer substance tritium to achieve what radium once did.

FACT BOX

Radium Girls

The First World War created a huge surge in demand for glow-in-the-dark instrument panels, watch dials and other devices. An American company called the US Radium Corporation launched a radium-powered paint which they called 'Undark'. This was used chiefly to treat military equipment, so soldiers could read their instruments under blackout conditions, but the company also made a version of the product for the domestic market

so users could illuminate their house numbers, light switches, or even, according to a company promotional leaflet, the backs of their slippers to make them easy to locate in the dark.

To keep up with demand, especially during the war, the company employed a large female workforce at their New Jersey factory. These women worked as dial-painters, each delicately applying radium-based paint to the hands and numbers of about 250 watch faces per day. The intricacy of the work meant that the camel-hair paintbrushes they were using required regular re-pointing, which the women, not recognising the danger, did with their lips and tongues. Consequently, they were inadvertently exposed to huge doses of radiation.

In an official report commissioned to investigate conditions at the factory, Harvard professor Cecil Drinker[*] wrote, 'Dust samples collected in the workroom from various locations and from chairs not used by the workers were all luminous in the dark room. Their hair, faces, hands, arms, necks, the dresses, the underclothes, even the corsets of the dial painters, were luminous. One of the girls showed luminous spots on her legs and thighs. The back of another was luminous almost to the waist.'

Not surprisingly, large numbers of these women subsequently developed horrific symptoms, including the loss of all of their teeth, jaw bones that crumbled like sponge

[*]   en.wikipedia.org/wiki/Radium_Girls

and disfiguring cancerous growths arising from the facial bones. The bones were the major manifestation of the radiation exposure, because radium behaves chemically a bit like calcium in the body, meaning that it tends to build up in bony tissue, which is why the teeth and jaws of the women were most affected.

It's not clear how many workers died as a result, but US Radium employed thousands. They were eventually taken to court on occupational health grounds and some of the women received small amounts of compensation, although most didn't live long enough to spend it. Their graves, however, remain radioactive to this day.

Science in the lap of luxury

As anyone who has ever kept a female dog knows, at certain times of the year she becomes the focus of affection for any male canine within sniffing distance. It's because she is 'on heat', which is another name for the mammalian oestrus, a time when animals advertise their fertility and attractiveness to the opposite sex. So does this happen to us? Prevailing wisdom says not and that evolution has abandoned the process in humans. After all, we're far too civilised to be obsessed with women's rear ends, aren't we? Apparently not, because thanks to a team of lap dancers in Albuquerque, it looks like the human oestrus is a myth no more.

Geoffrey Miller and his colleagues at the University of New Mexico[*] hypothesised that if women do subconsciously advertise their peak fertility in some subtle way, then men ought to find them most attractive at this time. If that's the case, then in a setting where men have to buy a

[*] *Evolution and Human Behavior* November 2007: Vol. 28, no. 6, pp. 375–81 DOI: 10.1016/j.evolhumbehav.2007.06.002

lady's attention, such as in a lap-dancing club, the woman's earnings from tips ought to peak in line with her fertility during her menstrual cycle. This fertile time is at ovulation, midway through their 28-day cycles between days 13 and 15. Critically, this is when, if a woman has sex, her prospects of pregnancy peak.

To test their theory, the New Mexico team recruited 18 local lap dancers and asked them to keep a daily tally of their earnings over a 60-day period. At the same time, the women recorded the stages of their menstrual cycles and whether or not they were using the oral contraceptive pill. The pill works by fooling a woman's body into believing that she is pregnant, which prevents ovulation and the other hormone changes that accompany the process, so it could have an effect on how attractive men find women to be.

The results of the study were 'sit up and take notice'-ably amazing. The earnings of the normally cycling (non-pill-using) lap dancers nearly doubled to US$350 per shift around the times when they ovulated and were therefore at their most fertile. Then, as they approached menstruation and their fertility fell, their earnings declined to a baseline of about US$200 per shift.

The pill users, by comparison, fared less well. They earned a flat average of US$200 per shift throughout their cycles. Why? Were the pill-using women just less attractive than their non-contraceptively compromised counterparts?

'No,' say the researchers, because both groups earned approximately the same amounts at the beginnings and ends of their cycles, indicating that the effect was not due to an overall difference in attractiveness. Instead, the results suggest that men can subconsciously detect when women are at their most fertile and judge them to be more attractive, valuable and, it would seem, worthy of more attention and larger tips at this time. As Miller points out, 'This is the first time that anyone has shown direct economic evidence for the existence and importance of oestrus in human females.'

It's not known what triggers the effect. It could be down to altered behaviour on the part of the fertile woman, such as chatting engagingly or wearing more provocative clothing and make-up, or it might be that chemical cues, like pheromones, are responsible. A strong possibility is that the sound of the woman's voice plays a part too. Indeed, the appropriately named Nathan Pipitone and his colleague Gordon Gallup, two researchers from the State University of New York at Albany,[*] have recently discovered that the menstrual cycle affects how attractive a woman sounds.

Evidence for this fertile form of conversation emerged when the scientists made recordings every week for one month of the voices of 30 female students counting to 10. About half of the students they studied were contraceptive pill users, the other half reported regular, natural menstrual cycles. The four recordings from each of the women were played back, in a random order, to a large panel of 'raters' consisting of equal numbers of male and female students who were asked to judge the attractiveness of the voices they were hearing.

When Pipitone and Gallup matched up the scores given to each woman with where she was in her menstrual cycle when each of the recordings was made, a pattern just like the lap dancers' tips emerged. The voices of non-pill-using females were rated as significantly more attractive at the times when the women were ovulating, and therefore most fertile, compared with other times

*   Evolution and Human Behavior July 2008: Vol. 29, no. 4, pp. 268–74
     DOI: 10.1016/j.evolhumbehav.2008.02.001

of the month. Amongst the pill users, on the other hand, the voice ratings barely changed across the month. There were also no differences between the verdicts of the male and female raters.

So how does this happen? It's probably down to the levels of the sex hormones oestrogen and progesterone. Scientists have found that lower levels of oestrogen and higher levels of progesterone, which occur towards the end of each monthly cycle, can cause the vocal cords to swell slightly, reducing voice pitch. And as lower-pitched voices, studies have shown, tend to be rated as less attractive than higher voices, this could explain the effect.

Regardless of how women broadcast their fertility – and researchers suspect that a combination of different signals is involved as a form of 'multiple messaging' – the effect does appear to be real. The bottom line would seem to be, ladies, if you want that new car at a knock-down price, consider ditching the pill, time your approach to coincide with day 14 and always head for the male sales assistant . . .

Women turn on testosterone (in men)

It's well known that women can influence each other's menstrual cycles using just the power of smell: ladies living together can synchronise their periods and, experimentally, women sniffing pads worn in the underarms of other females can alter their menstrual timings by up to two weeks. But if you still thought men were immune to the effect, then you've been myth-led – because males too, it now turns out, are equally sensitive to ladies' smells. In fact, research has shown that men exposed to the whiff of a woman experience a testosterone surge, although only on certain days of the month.

This came to light recently when two scientists at Florida State University, Saul Miller and Jon Maner,* asked a group of 37 heterosexual (and possibly 'metrosexual') young men to lend their noses in the name of science and sniff some ladies'

* *Psychological Science* February 2010: Vol. 21, no.2, pp. 276–83
DOI: 10.1177/0956797609357733

t-shirts. The men, who did not know the purpose of the study, were asked to smell t-shirts that had been worn by four non-pill-using women for three nights at points straddling the time when ovulation occurs in the middle of their cycles.

The men were also asked to smell a second set of t-shirts that had been worn by the same women, also for three nights, but this time towards the ends of their cycles (days 20 to 22), when fertility is low. A set of unworn t-shirts were also included as controls, and the women were asked, for the duration of the study, to use only neutral-smelling soaps and to avoid wearing perfumes or consuming foods with strong odours, like garlic or onions.

Before and after the men smelled the shirts, saliva samples were collected from each of them to measure their testosterone levels. They were also asked how 'pleasant' they found the shirt odour to be in each case. The results revealed that the men were rating the smells of the shirts worn around the time of ovulation (days 13 to 15) as much more pleasant. Moreover, the average post-sniff testosterone level was also significantly higher than when the men smelled control shirts or shirts worn by the women towards the ends of their cycles. So something in the shirts was peaking the men's sexual interests and provoking a libido-boosting burst of testosterone when the women were ovulating and most likely to conceive. Women, it seems, chemically augment their allure at certain times of the month.

This fits with the findings of Geoffrey Miller's lap dancer study (see previous chapter), in which scientists showed that lap dancers earn over 200% more in tips during their most fertile days. At the time of this earlier study, it wasn't clear whether the fiscal boost was just because the lap dancers were making themselves look more appealing or had sexier-sounding voices when they were at their most fertile. Now it looks like it was at least partly down to smell.

From an evolutionary standpoint, the results are exactly what we would expect: women exude some sort of pheromonal smell signal to broadcast their fertility to men who, in turn, as Miller and Maner put it, manifest 'mating-related behaviour'. By which, presumably, they mean a sudden inability to speak coherently coupled with the urge to drink too much, turn up the stereo to '11' and perform outlandish macho displays and skateboard stunts. And that's just the over-60s . . .

Solar flower power

Most people assume, correctly, that flowers look the way they do to attract insects that pollinate them. But that's not the whole story. Scientists have now discovered that plants have another 'trick up their leaves' to make themselves irresistible to even the most choosy insect – solar power.

Cambridge University's Beverley Glover and her colleagues[*] recently set up some fake flowers filled with a sugar solution, which they kept at different temperatures. Unleashing a team of bumblebees on their floral offerings, they watched as the insects visited the flowers to drink the surrogate 'nectar'. Very quickly, it became obvious that the bees were concentrating on the flowers with the warmest nectar. Just in case it was something to do with the colour of the fake flowers, the scientists also tried a different colour combination – and got the same result.

This proved that bees like their nectar hot, irrespective of the colour of the cup it's served in. But can flowers dish up hot beverages? 'Yes,' say the scientists, who have found that the surfaces of

*   *Nature* 3 August 2006: Vol. 442, p. 525 DOI: 10.1038/442525a

over 80% of flowers are covered with tiny conical-shaped cells which behave like microscopic lenses. These focus sunlight, warming up the flower and its nectar by several degrees. As the sun moves across the sky, the flower head moves too, keeping a fix on the sun like the dish of a radar tracking station.

But why would bees be interested in a warm drink anyway? Well, flight is an energy-hungry business, and by drinking hotter nectar the bee keeps warm for nothing. As Beverley Glover points out, 'For a bumblebee, we think it's about metabolic reward. They need the sugar from the flower to make energy to fly but they, like you on a cold day, might get more energy more quickly from a warm drink than a cold drink. It saves them from using their own energy to warm that nectar up if the flower's already providing it at a warmer temperature. The effect's also strongest at dawn and dusk and we know that bumblebees need extra help at these times of day when it's hard for them to get that big fat body off the ground.'

In the grand scheme of things, it's a clever way to save energy. And besides bees, other insects have also been spotted taking advantage of this natural plant solarium effect by basking in flowers to warm themselves up first thing in the morning. But hot nectar turns out not to be the whole of the story. The team suspected that the conical cells might also be giving bees a helping hand to clamber into flowers.

To test this idea,[*] they looked at how bees responded to a mutant snapdragon (*Antirrhinum*) plant which lacks these conical solar cells but looks and smells identical to a normal plant in all other respects. The lack of conical cells in these plants turns the flower surfaces into the floral equivalent of an icerink, making it hard to get a grip. Predictably, the bees universally turned their noses up at visiting the blooms, preferring the normal flowers instead. To prove that it was this genetic Teflon-effect that the bees were objecting to, the team then made casts of both the normal and the mutant petal surfaces using epoxy resin. These plastic flowers were then loaded with sugar solutions at identical temperatures and offered to the bees.

Incredibly, when these epoxy casts were presented horizontally, with the flowers lying

*   *Current Biology* 14 May 2009: Vol. 19, no. 11, pp. 948–53
DOI: 10.1016/j.cub.2009.04.051

flat and pointing upwards, the bees showed no preference in favour of either, but as soon as the epoxy blooms were hung vertically, so that the bees had to cling to the surface in order to drink the sugar solution, the bees showed a strong preference for the petal casts taken from the plants with conical cells. 'This is like bee velcro,' says Beverley Glover. 'Flowers have evolved these cells so that pollinators can get a grip on them.'

So flowers have killed two birds with one stone: by coming up with a conical cell covering, their petal surfaces can capture sunlight and serve up warm nectar, making the plant more metabolically attractive than a trip to the gym. But at the same time, these floral solar panels provide convenient stepping stones to help would-be pollinators get a grip.

The darkest surface: not all black and white

A simple soul like me could be forgiven for thinking that black does exactly what it says on the tin: it's black. But not anymore it seems, because scientists have recently shattered the myth that black is black by producing a substance 30 times darker. This new coating, which is the brainchild of Rensselaer Polytechnic researcher Pulickel Ajayan and his team,[*] consists of a sheet of nanotubes. These are very tiny but extraordinarily long molecular straws of carbon atoms, each about 5000 times thinner in diameter than a human hair.

They're grown as a film using a process called vapour deposition and the result resembles the nanoscale equivalent of a bamboo forest with the carbon tubes standing vertically, side by side, their tops entangled to make the surface appear irregularly corrugated. The individual nanotube

[*] *Nano Letters* February 2008: Vol. 8, no. 2, pp. 446–51 DOI: 10.1021/nl072369t

bamboos also vary in height and can be up to one millimetre tall. Prepared in this way, the result, says Ajayan, is a material that's very good at absorbing light but also very bad at reflecting it again, which is why it looks so dark. 'An ideal black material absorbs light at all wavelengths and all angles, basically.' The new material achieves this by allowing light to penetrate into the spaces between the nanotubes before being absorbed by the carbon in their walls. Even if a light ray manages to ricochet off one nanotube, it will still be absorbed by an adjacent one. 'So basically, light enters this material and it gets trapped.'

Compared with black paint or graphite, which reflect about 5–10% of the light that lands on them, this new nanosurface soaks up over 99.9% of the light that hits it, right across the visible spectrum, making it at least 100 times darker than even the average teenager's bedroom wall. The surface is also three times blacker than a nickel-phosphorus compound previously crowned the 'world's blackest substance', and 30 times darker than the carbon-based 'gold-standard black' held by the US National Institute of Standards and Technology.

But why is it useful? Apart from being an academic curiosity, and deserving of another entry in the *Guinness World Records* for the creator (his first was in 2007 for the 'world's smallest broom', a nanobrush with bristles 30-billionths of a metre in diameter), the discovery does have some potential spin-offs: it could, for instance, hold the key to invisibility cloaks to conceal the next generation of stealth bombers, or be used to create highly efficient solar cells that are capable of harvesting significantly more of the sun's energy than the 15–30% achieved by today's technology.

To pursue this, the Rensselaer team are now testing the substance to see whether it can soak up some of the other wavelengths of light beyond the visible spectrum, such as infrared, microwaves or possibly even X-rays and gamma rays. 'If you could make materials that would block these radiations, it could have serious applications for stealth and defence,' Ajayan points out, although it will be a little while yet before he can shed light on whether it actually works.

'BLACK' BOX

Another black myth is the name of the so-called 'black box' flight recorder carried on aeroplanes. Despite the misleading moniker, these devices, the first modern forms of which were developed in Australia in the 1960s and known as 'Red Eggs', aren't black at all but instead are painted bright orange to make them easy to locate in the event of a crash.

Although they are normally housed in an aircraft's tail section to maximise their chances of surviving an impact, today's flight recorders are nonetheless capable of withstanding head-on smashes at over 300 miles per hour and immersion in water to depths of over 6000 metres. To help crash detectives to recover them, they also fire off locator beacons to announce their positions.

Surprisingly, the idea for the flight data recorder first took off as early as the 1930s, in Europe. Two French aviation engineers eager to work out how to make their planes fly better, Francois Hussenot and Paul Beaudouin, developed a system that used mirrors to beam thin rays of light onto strips of slowly advancing photographic film. By adjusting the angle of the mirror, the light beam position could be altered to record parameters like altitude and speed. Developing the film would then reveal a line charting the course of a flight. Although no one knows for sure, this could be the origin of the name 'black box', because it would have been necessary to keep the film in the dark to avoid accidental exposure and hence loss of the flight recording.

These days, rather than the eight metres of film used in Hussenot and Beaudouin's recorders, modern devices use solid state electronic systems to log up to 25 hours of flight data, including any movements or adjustments of the aircraft controls, the performance of the major systems as well as communications amongst the flight crew and with air traffic controllers. This provides a forensic record to help investigators trace the causes of crashes but, increasingly, the information is also being used by manufacturers and developers to work out how to fly planes at maximum efficiency to minimise fuel costs and reduce greenhouse gas emissions.

Painful genes

All in the mind, mind over matter, push through the pain, high pain threshold, no pain no gain. Ask a muscle man, marathon runner or even a woman in labour what pain is all about and they'll often answer 'willpower'. But now a family of feted firewalkers, impalers and stunt daredevils – who can't feel pain at all – have revealed to researchers that pain is as much down to DNA as it is to mental fortitude.

Cambridge-based scientist Geoff Woods[*] was in Pakistan when he first heard about several families who appeared to be able to weather pain with impunity. To entertain their friends, for instance, affected children would go as far as to jump off roofs or even plunge their hands into boiling liquids, often sustaining horrible injuries in the process, but without apparent concern. The reason? Despite being neurologically otherwise normal, these individuals could not tell when something should be painful. They could tell hot from cold and smooth from rough, they were ticklish and they enjoyed curry, although in the

* *Nature* 14 December 2006: Vol. 444, pp. 894–98
DOI: 10.1038/nature05413

latter case it didn't burn them in the same way that it would a normal person.

Attractive as a pain-free existence sounds, it's actually bad news because, as the affected individuals sadly showed, not being able to experience pain means you can't always tell when you're doing real harm to yourself, inadvertently or otherwise. More unfortunate still, some might argue, the affect does not extend to numbing the excruciating sound of James Blunt.

However, there is a plus side, at least genetically speaking, because the bizarre all-over anaesthesic affect manifest amongst the members of these Pakistani families was found to be passing directly from parents to offspring. This sent a molecular-biological shiver down Woods' spine, because he realised that it meant it could be down to a single dodgy gene. Finding this gene might therefore unlock some of the secrets of 'nociception', including new ways to hit pain harder where it hurts.

Using DNA samples collected from six of the Pakistani family members, and comparing affected with unaffected individuals, Woods and a team of researchers performed the genetic equivalent of combing through a genome-sized haystack to find a very small needle. Their endeavours were rewarded when they successfully homed in on a region of chromosome number two that contained the genetic cause of the Pakistani families' insensitivity to agony. The gene they uncovered is called SCN9A. Normally it codes for a channel, resembling a tiny pore, which allows sodium to enter and excite pain-signalling nerve cells in response to painful experiences. Among members of the affected families, however, the gene contained a mutation that prevented it from functioning correctly. As a result, their nerve cells were effectively deaf to the screams of painful events going on around them.

The story doesn't end there, because then the team began to wonder whether other forms of the gene might exist, accounting for why some people are more stoical than others. And, thanks to a bunch of people with backache, the answer, it appears, is yes! The researchers discovered this[*] by comparing the pain scores reported by 578 arthritis patients with the severity of their disease as it appeared in X-rays of their joints. Some patients, the team found, were reporting much more pain than others, despite having

* _PNAS_ 16 March 2010: Vol. 107, no. 11, pp. 5148–53 DOI: 10.1073/pnas.0913181107

Stripping Down Science

arthritis of a similar severity. To find out why, they matched up the patients' pain scores with the DNA sequences of their SCN9A genes. This led the team to identify two variants of the gene in the patients, a rarer A form and a more common G form. On average, they found patients carrying the A variant tended to report more severe pain than patients carrying the G form of the gene.

To confirm the findings, the researchers repeated the study on 179 patients with lumbar back pain, with similar results. They also subjected a group of female volunteers to a range of painful stimuli, again demonstrating that individuals carrying the A form of the gene were more pain-sensitive. To find out why this genetic variation was having this effect, they then expressed the SCN9A gene in cultured HEK293 cells, which have nerve-like properties. In these cells the A and G forms of the gene had subtly different electrical effects, sufficient to explain the increased sensitivity of carriers of the A form to painful stimuli.

So it seems that sensitivity to pain, and an individual's threshold for recourse to the pill packet, is as much under genetic influence as it is an exercise in tough-mindedness! More seriously, as and his colleagues point out, understanding how to modify the functions of genes like SCN9A will lead to better analgesics with fewer side effects, as well as the identification of those who have more specific painkilling requirements.

FACT BOX

Other ways genes can provoke a deleterious lifestyle

As well as determining physical fortitude and pain-sensitivity, the genes we carry are also instrumental in controlling behaviour, because they dictate how the brain develops and wires itself together. This means that now, thanks to the human genome project, scientists can begin to discover how different combinations of genes, and variations within individual genes themselves, can affect the kinds of lives we lead and to what diseases we're likely to succumb.

One area that is attracting a lot of interest is nicotine addiction. Apart from caffeine and alcohol, tobacco is one of the world's most popular legal drugs, but it's also directly responsible, according to the World Health Organisation (WHO), for causing the premature deaths of up to five million people every year. Most of these deaths are from heart disease, as well as lung cancers, chronic bronchitis and emphysema.

Amongst the smoking population, some people appear to be far more genetically vulnerable to developing one or more of these disorders than others, and there are also individuals who are genetically more prone to become hooked on tobacco in the first place. But can we tell who these people are, so that we can warn them?

According to research being carried out at the University of Auckland by chest specialist Dr Rob Young and his colleagues,[*] the answer is 'yes, definitely'. To reach this verdict, they've been comparing the genetic make-up of large numbers of individuals who have smoked for many years yet have not developed any chest diseases, with a second group of patients with similar histories of tobacco exposure but who have developed chest problems. This

[*]   As told to Dr Chris Smith.

has enabled the researchers to home in on 19 different genes that together can be used to predict an individual's risk of developing a smoking-related disorder. This means that those at high risk could now be warned ahead of time – while their lungs are still healthy – to stop smoking before it's too late.

The evidence is that more than three-quarters of smokers want to kick the habit, but only about 5–10% are successful in any given year. Young hopes that removing some of the uncertainty from the situation and highlighting the risks might help to motivate quitters to be more successful. 'We think that there are people out there who, given the information, might be prompted to be more proactive and engage in quitting activities with greater vigour and greater success, and there is data to support this.'

But what about getting hooked in the first place? Well, that's down to genes too. Jacqueline Vink, from VU University in Amsterdam,* screened the DNA of 3500

* *American Journal of Human Genetics* 5 March 2009;
   Vol. 84, no. 3, pp. 367–79 DOI: 10.1016/j.ajhg.2009.02.001

people, including smokers, past-smokers and nonsmokers, to look for genetic sequences that cropped up more often amongst the smokers or previous smokers than nonsmokers.

A number of DNA hotspots emerged from the analysis, which she then checked in a further three groups of people containing about 400, 6000 and 1600 participants respectively. The result was the identification of a clutch of genes that had never previously been linked to smoking, which included receptors for the excitatory nerve transmitter chemical glutamate, genes that control the transport of chemicals into nerve cells and genes encoding adhesion molecules to help cells to link together. The motivation for carrying out the study was straightforward: 'Identification of genes underlying the vulnerability to smoking might help identify more effective prevention strategies and thus diminish smoking-related morbidity.'

That said, according to Mark Twain, 'To stop smoking is the easiest thing I ever did. I ought to know; I've done it a thousand times.'

Veins contain blue blood . . . don't they?

An urban legend, often bandied about, is that blood changes to a blue colour when oxygen is removed from it, so veins, which carry deoxygenated blood back to the heart, look blue beneath the skin.

The myth owes its origins to the Spanish upper classes who championed 'sangre azul' (blue blood) as a sign of pure breeding. Their thinking worked along the lines that blue veins were only visible beneath the pale skin of someone with a bloodline untainted by the darker-skinned Moors, who had controlled large parts of the country in years gone by. Unfortunately, the expression has since led to some serious scientific disinformation.

Blood is a rich red colour because it contains haemoglobin, a protein which carries oxygen from the lungs to the tissues. Haemoglobin is actually four proteins stuck together, known as a tetramer, comprising two alpha-haemoglobin and two beta-haemoglobin molecules, each with an iron atom at the centre. It is this iron which gives haemoglobin its colour. Each molecule of haemoglobin can bind four molecules of oxygen, and when the oxygen unites with the haemoglobin it changes its shape and its light absorbency so that it reflects relatively more red light. So oxygen-rich blood in arteries, on its way to our tissues, looks much redder than venous blood.

In hospital, doctors sometimes collect arterial blood samples to see how much oxygen is present and hence how well the lungs are working. In a well-oxygenated person the blood is a bright brick red. If the person is very short of oxygen, or if you miss the artery and instead collect the sample from an adjacent vein, the blood appears much darker and is a red-black maroon colour … but still, not blue. So while it seems reasonable to assume that the colour of veins is partly down to the colour of blood they contain, that's not the whole story. The give away is that superficial veins (those just below the skin surface) don't look blue at all. Anyone with 'thread veins' on their legs or face knows only too well that they are red. So what's going on?

Some light was shed on this problem in recent years by Alwin Kienle at the Institut fur Lasertechnologien in der Medizin, in Ulm,

Germany.* He set up a model vein comprising a blood-filled tube suspended in a milky solution designed to mimic the optical properties of the skin. By moving the blood vessel up and down in the solution, to simulate deeper or more superficial veins, it was possible to measure the different wavelengths of light being reflected and hence work out why deeper vessels look blue. Sure enough, when he immersed his surrogate vessel at greater depths, it ceased to appear red and instead took on a blue hue without actually changing colour at all.

The reason for this colourful conundrum, Kienle found, is due to a combination of the way the brain decodes colours, together with the fact that different tissues absorb some wavelengths (colours) of light more than others. In general, blue light penetrates into the skin much less well than red light. So if a vein is near the surface of the skin, as in a thread vein, almost all of the blue light hitting the tissue is going to be absorbed. This means that relatively more red light will be reflected back, making the vein look red overall.

So far so good. But what about a deeper-situated

*   *Applied Optics* 10 December 1996: Vol. 35, no. 7, pp. 1151
DOI: 10.1364/AO.35.001151

vessel? This, Alwin Kienle found, is where the way that the brain processes colour comes in. Even though red light penetrates further through skin than blue light, for veins sited more deeply within the skin the amount of red light reaching them begins to dwindle. At the same time, the deoxygenated blood they contain soaks up some more of the incoming red light, which is why venous blood is a darker red. This means that light coming back from the vein is slightly less red – in fact, it's a bit more purple – than light reflected back from the tissues adjacent to the vein. Because the brain processes colours in a relative way, if something purple is viewed next to something red, it ends up looking blue, hence the illusion and the myth.

Despite blue blood being a myth for humans, for some animals, including lobsters, crabs, shrimps and other crustacea, it's a reality. The reason for the colour change is that these creatures use copper, instead of iron, in their equivalent of haemoglobin, which is a protein called haemocyanin. For them, blood is blue when it is oxygenated and gradually loses its colour as the oxygen is removed. But the real hippies of the haemoglobin world are definitely the brachiopods, sipunculids, priapulids and magelona (a type of worm). These animals have a blood pigment called hemerythrin, which is a bright violet pink when it's charged up with oxygen.

Stripping Down Science

Vile-din?
Certainly not!

Ask anyone who made the world's best violins and they'll inevitably answer 'Stradivari'. But science is beginning to undermine the reputation of this great instrument maker who, it seems, might owe at least part of his success to an attempt at chemical pest control, rather than just his craftsmanship.

Antonio Stradivari was born in 1644 and lived in Cremona, a city in north-west Italy. He set himself up as an instrument maker in the 1680s but his 'golden period', during which he is believed to have produced some of his best instruments, didn't come until the 1700s, by which time he was over 70 years old. About 600 of his instruments are thought to survive today and, in good condition, they are each worth at least US$5 million. The hefty price tag reflects the fact that not only are they 300 years old, they're thought to be genuinely unrivalled in terms of the quality and purity of the sound they produce. Effectively they're the Rolls-Royce Silver Ghosts of the musical world.

Not surprisingly, very few owners are willing to donate their instruments 'in the name of science' to help researchers find out why they are so special. But it's been a lifetime ambition of Hungarian-born scientist, musician and violin maker Joseph Nagyvary, who's also an emeritus professor of biochemistry at Texas A&M University,[*] to do just that. Now, thanks to some tiny wood fragments donated by restorers working on these violins, he thinks he knows the answer.

Nagyvary used a technique called infrared spectroscopy to dissect the chemical structure of the wood in the fragments. He then compared it with similar samples collected from an old English and an old French instrument dating from the same period. The results were striking. The trace from the Stradivarius was very different from the other European instruments. It showed signs of having been chemically brutalised. The amount of lignin in the wood was reduced, and the hemi-cellulose, which acts like a molecular bridge holding the wood together, was greatly damaged. This would dramatically alter the

* _Nature_ 30 November 2006: Vol. 444, p. 565 DOI: 10.1038/444565a

resonant properties of the wood and change its acoustics, accounting for the distinctive sound said to single out these instruments.

But what could have caused this degradation in the wood? In an attempt to reproduce the effect, Nagyvary tried boiling and even baking samples of modern wood, but the treatment wasn't harsh enough. Instead, it seems Stradivari, or the carpenter who supplied him, must have resorted to chemical means, probably in the form of copper and iron salts, which are strongly oxidising and could conceivably have damaged the wood in this way. To find out exactly what chemicals they must have used will require access to more wood fragments, which could take some time. 'These samples are hard to get,' Nagyvary says. 'You cannot approach Itzhak Perlman and ask him to give you a chunk of his Stradivarius for analysis.'

But why chemically massacre your future instrument anyway? Nagyvary thinks the answer is all down to a primitive attempt at preservation. 'I am a heretic in this regard. I really don't think that Stradivari did this for acoustical purposes. I think that was a rather routine process around that time, in Cremona, where most woodworkers had to preserve their wood against the woodworm. Stradivari was a marvellous craftsman,' Nagyvary observes, 'but the magnificent sound of his instruments is a lucky accident.'

When does a fruit fly not smell like a fruit fly? Answer: when it smells like a mosquito

You're having me on! I hear you exclaim. But no, mad as this might sound, it's no myth. With the magic of modern science it's possible to give a laboratory fruit fly the smelling ability of a mosquito, which is exactly what a group of scientists did recently to work out how these bloodthirsty winged menaces hunt us down for dinner.

Mosquitoes are universally acknowledged as the most dangerous animals on earth owing to the number of deaths they cause by spreading diseases like malaria, dengue and yellow fever, which together run to hundreds of millions of cases per year. Scientists and doctors are therefore very eager to track down how it is that mosquitoes home in on us, and what attracts them to humans in the first place, because if we can understand how they're doing this then we can come up with better repellents. At the moment, substances like DEET (diethyl toluamide) are produced by chemical trial and error, but by knowing exactly how a mosquito identifies its next meal, it ought to be possible to produce more macho molecules tailor-made to turn a hungry mosquito into an anorexic.

Fundamental to the mosquito's human-tracking ability is its olfactory arsenal. The antennae that project from its head are covered in receptor molecules, which resemble miniature chemical docking stations that are each wired up to an individual nerve that connects to the animal's brain. Different types of receptors are specialised for picking up different types of odour molecules, which tend to have specific chemical shapes or structures. So when one of these odour chemicals bumps into a receptor that is the right shape to recognise it, the receptor is activated and fires off impulses down the adjacent nerve, signalling a 'hit'. The brain then adds together all of the incoming information to build up a picture of what the world smells like and from which direction certain odours are arriving.

This sounds simple enough but, by studying the mosquito genome, scientists have found the genetic recipes for more than 70 of these different receptors. They've also found that many odour chemicals can activate more than one type of receptor at once, which makes it very difficult to understand exactly which odours are detected by which receptors and therefore how to make an insect repellent to best block them.

Now enter Yale researcher Allison Carey.[*] Using a family of mutant fruit flies in which one of the groups of nerves in their antennae are devoid of odorant receptors, she added to these anosmic neurons, one at a time, the genes for 72 different mosquito odorant receptors. This was the molecular equivalent of stitching a dog's nose onto a human, because it bestowed on the resulting flies the ability to smell whatever chemicals each of the individual genes normally enable mosquitoes to detect. By recording the electrical activity from the individual nerves inside the fruit fly antennae as the insects were exposed to 100 different chemicals and odours in turn, including substances known to be produced by

*   *Nature* 4 March 2010: Vol. 464, pp. 66–71 DOI: 10.1038/nature08834

humans and the bacteria that live on human skin, it was possible to work out what contribution each gene makes to the smell-detecting repertoire of a mosquito.

This means that researchers now know which classes of chemicals mosquitoes can respond to and which ones are likely to be key targets in developing novel repellents. For instance, 27 of the receptors studied responded particularly strongly to compounds found in human sweat. Now they've been identified, research is focusing heavily on how these receptors work, how they interact with the odour molecules they detect and how best to block them or, paradoxically, activate them.

'We're screening for compounds that interact with these receptors,' says John Carlson, one of the other scientists involved in the study. 'Compounds that jam these receptors could impair the ability of mosquitoes to find us. Compounds that excite these receptors could help to lure mosquitoes into traps or repel them. The best lures or repellents may be cocktails of multiple compounds. The world desperately needs new ways of controlling these mosquitoes, ways that are effective, inexpensive, and environmentally friendly.'

FACT BOX

Turning mosquitoes into flying vaccinators

Nuisances as mosquitoes are, and desperate as scientists are to disable them, researchers have nonetheless also been sizing them up as mobile hypodermics capable of delivering a flying vaccination service. The work is based on the premise that, every time they take a blood meal, female mosquitoes first inject their saliva, containing anticoagulants and immune-evading agents, around the blood vessel puncture site. This is what provokes the itchy inflammatory aftermath but is also responsible for transmitting infectious agents like malaria, which the mosquito regurgitates into the wound when it feeds.

Shigeto Yoshida, from Jichi Medical University in Japan,[*] reasoned that it ought to be possible to exploit this unpleasant aspect of the insect's behaviour in a beneficial way.

* _Insect Molecular Biology_ June 2010: Vol. 19, no. 3, pp. 391–98 DOI: 10.1111/j.1365-2583.2010.01000.x

He set about genetically modifying *Anopheles* mosquitoes to make them produce in their saliva a protein called SP-15, which is critical for the spread of another major disease-causing parasite called *Leishmania*. Mice bitten repeatedly by these modified mosquitoes developed antibodies to SP-15, which other researchers have shown can protect against *Leishmania* transmission. 'Following bites, protective immune responses are induced, just like conventional vaccination but with no pain and no cost,' says Yoshida. 'What's more, continuous exposure to bites will maintain high levels of protective immunity, through natural boosting, for a lifetime.'

The next step will be to test whether mice vaccinated by these mosquitoes really can be protected from *Leishmania* infection in future. The odds are that it should work, because the same SP-15 protein has been successfully tested as an experimental vaccine previously. But whether this flying vaccinator technology will take off in general and for other types of vaccines is another matter. Some may feel slightly stung by the idea of a natural and uncontrolled vaccination system delivering unmetered drug dosages and boosters indiscriminately.

This sentiment is pre-empted by Yoshida himself in his paper describing the work in the journal *Insect Molecular Biology*: 'The concept of a "flying vaccinator" transgenic mosquito is not likely to be a practicable method of disease control, because "flying vaccinator" is an unacceptable way to deliver vaccine without issues of dosage and informed consent against current vaccine programs. These difficulties are more complicated by the issues of public acceptance to release of transgenic mosquitoes.'

Still, it might be one to have in the bag for when times get really tough!

FACT BOX

Malaria in children: a case of mistaken identity

Mosquitoes are directly responsible for causing over 500 million cases of malaria per year, the annual death toll from which is over one million, mostly young babies and children aged between six months and three years in sub-Saharan Africa. Why does this disease hit this age group much harder than any other? The answer, it turns out, is a case of mistaking friend and foe.

Christopher King, a researcher from Case Western Reserve University in the US,[*] followed up 586 babies born in Kenya from the time they were born until the age of three. He collected umbilical cord blood samples from the babies and also took a specimen of blood from each of the babies' mothers. Tests showed that some of the mothers were infected with malaria at the time when they

[*]   *PLoS Medicine* July 2009: Vol. 6, no. 7
      DOI: 10.1371/journal.pmed.1000116

gave birth, suggesting that the baby might also have been exposed to malarial antigens – chemical markers made by the parasite – which would have been circulating in the mother's bloodstream.

King and his colleagues therefore suspected that this might be causing the babies' immune systems to develop what is known as 'tolerance' to malaria. Tolerance is a process by which the immune system normally learns what it should befriend (and ignore) and what it should attack. If the malaria parasite is present when the baby is developing, the team reasoned, the baby's immune system might be fooled into thinking that the parasite is a normal part of the body and so ignore it.

In keeping with this theory, when the team mixed malaria antigens with white blood cells from babies whose mothers were infected with malaria at the time of delivery, the babies' cells reacted only very weakly. But white cells from babies who were not born to infected mothers, on the other hand, showed vigorous reactions and pumped out large amounts of inflammatory hormones.

This suggests, say the team, that if a mother is infected with malaria when she is pregnant, the baby's immune system is misled into becoming tolerant to the parasite rather than attacking it. This could have serious implications for the development of a successful vaccine, because the children who are most at risk are those who are already tolerant to malaria and therefore won't make a very powerful immune response to a vaccine.

So why are children under six months less affected? This is because babies are initially protected from malaria by antibodies which are present in breast milk and are also added to the baby from the mother's bloodstream during the final phase of pregnancy. These antibodies persist in the baby's circulation for about six months, but once they are gone the child becomes vulnerable. And if the child is one that has developed tolerance to malaria, then it will develop far more severe malaria disease with a correspondingly increased risk of dying.

A saving grace is that researchers at least now know about this previously unappreciated part of the malarial parasitic puzzle. But as yet, the world is still waiting for an effective vaccine for one of the most common killers.

Pure as snow?
Not so!

We often think of snow as a paragon of purity; it can transform even an ugly industrial landscape into a soothing sight that's easy on the eye. But snow's pristine image is something of a myth, because beneath its sparkling exterior lurks a dirty secret, the full scale of which is only now becoming apparent.

Analysing samples of freshly fallen snow collected from a range of locations around the world, Louisiana State University scientist Brent Christner and his colleagues[*] found that the snow contained large numbers of bacteria which, it turned out, had quite literally come down in the last shower. Why were these bacteria turning up in pristine snow? Well, the results suggest that they were most likely hitching a ride around the world in snow clouds, and to help them to do that they've evolved a clever way to manipulate the formation of ice crystals.

When water-soaked air rises up into the

[*]   *Science* 29 February 2008: Vol. 319, no. 5867, p. 1214 DOI: 10.1126/
      science.1149757

atmosphere, it cools until it eventually reaches a temperature at which the water can no longer remain as a vapour and so it begins to form droplets of water or ice. The droplets usually gather around particles of soot, pollen, dirt and even dandruff, which provide surfaces called nucleation sites on which the water can condense. But these structures only condense water efficiently at very low temperatures, below minus 10 degrees Celcius, and this is where the cloud-riding bacteria come in. The surfaces of their cells are peppered with a large protein which contains chemical groups that love to latch onto water molecules.

These chemical groups are arranged at just the right spacing and orientation so that they mimic the crystal structure of ice. Once water molecules have locked on, as soon as the temperature dips just below zero, they're already in the ideal arrangement for freezing, which means that ice can form around the bacterium at much higher temperatures than it would do normally.

So why should a bacterium want to seal itself into the middle of a microbial ice cube? Probably because these bugs, called *Pseudomonas syringae*, are a family of plant pathogens that use frost to make a forced entry into the leaves and stems of their hosts. By triggering the formation of ice on a plant's surfaces, the bacteria can use the jagged crystals to punch holes into the tissue. This microbial equivalent of a ramraid allows the bacteria to penetrate the plant and soak up nutrients. This trick was first discovered in the 1970s, and very quickly researchers realised that if the bacteria were mixed with water and sprayed onto a near-naked ski-run they could produce a perfectly passable piste under the right conditions. In fact, they're marketed today as 'Snomax', although they come freeze-dried, rather than ready frozen.

But what are they doing in clouds? Well, although these bugs are found everywhere in nature, until recently no one had appreciated how they were getting around. Now scientists suspect that strong winds blow them from plants up into the air and carry them for tens or even hundreds of kilometres. Then, as they go, they use their clever chemistry to trigger clouds to form around them and descend with the ensuing shower onto a new patch of greenery below. So where this piece of science is concerned, it seems that it never rains until it pours.

How does a heart come by its own arteries?

A question that's been circulating for over 100 years is where do the heart's own blood vessels – the two coronary arteries – come from? These vessels follow a characteristic path from the aorta, the body's main artery, around the two sides of the heart, branching as they go to supply each part of the muscle. Owing to this pattern, embryologists and anatomists thought that these arteries grew out from the aorta and across the heart's surface as the organ was developing. But this turns out to be a blood-vessel-bustingly-big myth, because new research has shown that the coronaries actually come from a large vein outside the heart. This matters because if we can now work out what the signals are that control the process, it might be possible in the future to reactivate them and so make broken hearts mend themselves.

The new discovery was made by a scientist at the Howard Hughes Memorial Institute in

America called Kristy Red-Horse.[*] By using a genetic label to trace, in mice, the movements of the cells that become blood vessels around the heart, she was able to show that the coronary arteries were springing up from a very unusual source, nowhere near the aorta. Instead, the first thing that happens, she found, is that very early during development a cluster of cells migrates onto the rear surface of the developing heart from a large vein passing behind it called the sinus venosus, which carries blood in the embryo back to the developing heart.

These cells, which are known as endothelial cells and normally form the inner linings of blood vessels, then migrate around and plumb themselves into the heart's main outflow tract – the future aorta – which is situated in the centre at the top of the developing organ. From there, the cells make their way down the left side and the front of the heart, laying the foundations for what will become the future left coronary arteries. The cells also travel along a strip down the back of the heart to form the right coronary artery. As they move over the organ's surface, the cells invade the

*   Nature 25 March 2010: Vol. 464, pp. 549–53
     DOI: 10.1038/nature08873

muscle tissue, producing capillaries to deliver the oxygenated blood flowing along the arteries to the actively beating cells. They also give rise to veins to return the spent blood to the heart's right atrium so that it can be reoxygenated by the lungs.

That all sounds relatively straightforward, but a key question is where do these artery-producing cells come from in the first place, and what directs them to achieve this impressive feat of cardiac plumbing? The answers to these questions might hold the key to growing new blood vessels to bypass coronaries clogged by cholesterol in patients with heart disease.

By carefully examining the sinus venosus, Kristy Red-Horse and her colleagues were able to spot the place where the cells were originating from the lining of this vein. After they appeared, rather like miniature moles, the cells burrowed through the wall of the sinus venosus to emerge onto the rear surface of the heart. This seems to occur in response to a chemical signal produced by the adjacent developing heart tissue, because when pieces of either the sinus venosus or the heart tissue were cultured in a dish in isolation, no vessel-forming cells appeared. But when the two were brought close together, coronary critical mass was achieved and the cells began to grow and spread.

What makes this happen, and the nature of the other signals that guide the artery-producing cells across the heart surface, isn't known yet. But the researchers did find that the endothelial cells leaving the sinus venosus all turned off various genes that are associated with vein tissue and then, as soon as they entered the heart tissue, turned on a different combination of genes that are normally expressed in arteries.

This shows that vein cells can effectively be reprogrammed to build arteries and that the heart muscle must produce signals to direct and control this process. Now the race is on to work out what they are. As one of the team, Mark Krasnow, puts it, 'If we can learn how to reprogram cells to build a new coronary artery just like the original, bypass grafts could last the rest of a lifetime.'

FACT BOX

Heart disease and repair

Heart disease kills about one person in every three, making it the most common cause of death worldwide. Heart attacks happen when the blood vessels that supply the heart muscle – the coronary arteries – become blocked, usually through a build-up of a fatty substance called atheroma. One way to deal with this problem is to carry out a bypass operation, where a piece of blood vessel – usually a muscular vein from the leg called the long saphenous vein – is used to route blood around the blocked part of the vessel, restoring flow downstream.

Although this sort of surgery is still occasionally performed, these days doctors are more likely to carry out a procedure called angioplasty and stenting. This involves the threading of a thin tube into one of the arteries in the leg and from there up to the heart and into the blocked coronary artery. Using X-rays, the tip of the tube is positioned over the blockage and a small balloon inflated for a few seconds to open up the narrowed area. To prevent the vessel closing up again, a tiny metal cage called a stent is then inserted into the treated spot to prop open the artery walls.

These treatments have revolutionised the management of coronary artery disease, but unfortunately they are less useful for patients who have already had a heart attack, which has led to the death of a patch of heart muscle. In humans, the damaged area never recovers its pumping ability and is instead replaced with fibrous scar tissue, which can lead to heart failure and the formation of blood clots.

Not all species behave like this though. In fact, some, like the tiny zebra fish, can regenerate a sizeable chunk of their heart if it is damaged, leading scientists to wonder if it might be possible to recapitulate the trick in humans. But how these animals were doing this remained an unsolved cardiac conundrum, at least until very recently.

One idea was that the fish had some kind of stem cell lurking in their heart tissue, ready to spring into action and regenerate new muscle if anything went wrong. This turns out not to be true, however, because Barcelona-based researcher Chris Jopling and Howard Hughes Medical Institute scientist Kazu Kikuchi[*] have found that heart muscle cells themselves have the necessary know-how to carry out even fairly significant cardiac DIY. The researchers showed this by genetically tagging a group of fish with a coloured protein to label up just the heart muscle cells. They then removed 20% of the ventricle tissue and watched to see what happened. If the heart muscle cells grew and divided to replace the lost tissue, then the repair should comprise cells all carrying the coloured genetic marker. But if an unlabelled stem cell was responsible, then the repair would also be unlabelled and the new muscle cells would be colourless.

In fact, what they saw was that the injury provoked nearby surviving mature heart cells to temporarily shed their muscle-cell-like appearance, resembling stem cells instead.

[*] *Nature* 25 March 2010: Vol. 464, pp. 601–05 DOI: 10.1038/ nature08804; *Nature* 25 March 2010: Vol. 464, pp. 606–09 DOI: 10.1038/nature08899

These cells then began to divide, producing replacement muscle cells, all of which bore the coloured genetic marker. The regenerating cells also turned on another gene called gata4, which is normally active during embryonic development of the heart, suggesting that the heart muscle repairs by reactivating the same genetic program that was used to build the heart in the first place inside the embryo. Working out how to make a damaged human heart do the same is therefore a major priority for researchers worldwide.

Is the sea really blue?

A commonly held myth is that the sea is blue because it reflects the colour of the sky. In fact, it's the strange chemistry of water that gives the sea its colour, and has also led some underwater species to evolve the fish equivalent of an infrared spy camera.

Water is one of the most abundant molecules on the earth. Most of it arrived here from space in the form of ice aboard comets that have rained down on the earth over the four and a half billion years since the planet first formed. Geologists have calculated that there's almost one and a half billion cubic kilometres of water on earth, and without it life would probably never have got off the ground, or even off the seabed, in the first place.

That's because water has some unusual chemical properties. Each molecule, formula H_2O, resembles a tiny boomerang with the oxygen atom at the apex and the hydrogen atoms at each of the tips. This structure soaks up light in the part of the spectrum that we feel as heat, or infrared, which is why water vapour in the atmosphere is actually a more powerful greenhouse gas than carbon dioxide. In this respect, it helps to keep the planet warm.

But in the ocean, where water molecules are surrounded by many other water molecules, a phenomenon called hydrogen bonding occurs. Put simply, these are weak 'intermolecular forces' that arise because water is what's known as a 'polarised molecule'. The central oxygen (O) of the H_2O pulls the electrons of the two hydrogens towards itself, making the hydrogens slightly positively charged and the oxygen slightly negative. Since unlike charges attract, a hydrogen on one molecule will be attracted to the oxygen of another adjacent water molecule. This makes water sticky and accounts for many of its wonderful life-sustaining properties.

The consequence of this hydrogen bonding is that it makes the molecules 'stiffer', so when light hits water, instead of absorbing just in the infrared, it begins to absorb more strongly at visible red wavelengths too. Since more red wavelengths are being soaked up, leaving behind relatively more blue light, the water looks blue. And the deeper you go, the greater the amount of red light that has been removed, so the 'bluer' everything looks. In fact, at depths over a few hundred metres, only a very narrow range of visible wavelengths can make it through, and these are entirely at the blue end of the spectrum. As a result, many of the marine species that live at these depths lack the necessary chemicals in their eyes to even 'see' red light, leading scientists to believe that this was true of all these animals.

But therein lies another myth, because marine biologists recently stumbled upon some fish species that have turned this disability into a weapon. The deep-sea stomiid dragon fishes *Malacosteus*, *Aristostomias* and *Pachystomias* have evolved an organ that sits beneath their eyes and pumps out red light. The fish have also altered their retinae, adding a number of extra light-detecting chemicals. These adaptations mean that the fish can see both their own and the red light produced by other members of the same species, while other fish remain oblivious to the fact that they are being 'spotlighted' and possibly eyed up for dinner. They are, quite literally, left in the dark. Apart from hunting, however, this covert communications system also makes light work of seeing potential predators before they see you, and in finding a mate, which puts a whole new spin on the idea of a red light district . . .

*   *Marine Biology* 2005: Vol. 148, pp. 383–94 DOI: 10.1007/s00227-005-0085-3

Stripping Down Science

Bee careful: why bees are nature's risk assessors

Bees are universally viewed as shining examples of hard workers who also dance for their dinner: when one bee finds a fruitful food source, just like motorists tipping each other off about low-cost fuel, the discoverer returns to the nest where she performs an intricate 'waggle dance' to inform hive-mates where to fly so they too can get nectar and pollen on the cheap. Previously, scientists thought that was the 'bee-all' and end-all of the story, but now it turns out to be something of a myth, because researchers have discovered that there's a flip side to the waggle dance: a flight-deterrent manoeuvre that takes the form of a headbutt.

Originally, bee researchers believed this behaviour was a begging call for food. But James Nieh, a San Diego-based scientist,[*] has scratched that claim by watching the behaviour of bees

[*]   *Current Biology* 11 February 2010: Vol. 20, no. 4, pp. 310–15
     DOI: 10.1016/j.cub.2009.12.060

returning to their hives after being attacked during foraging flights by members of rival bee colonies. He noticed that if a bee that had recently been mugged by rivals during one of its food-gathering flights spotted another bee using its waggle dance to recruit other workers to forage in the same potentially lethal location, the bruised bee would start headbanging the dancer to make her stop.

Slowed down to make it easier to see, the movement actually consists of the sender of the signal butting her head into the recipient and then vibrating her body 380 times per second for about one-tenth of a second. This neutralises the dance and alerts the performing bee to the fact that she could be leading the hive off in a potentially dangerous direction, so she stops.

Moreover, the bees seemed to have a greater gift for risk assessment than an OH&S inspector. Nieh noticed that the number of times the stop signal was issued by an individual was proportional to the level of danger it had faced previously when foraging. Survivors of 'significant aggression' – presumably the bee equivalent of 'did you spill my beer?' – upped their rate of warnings 40-fold, while others that had their legs mechanically pinched with tweezers to simulate a bite from another bee delivered 88-fold more warnings. 'Butt' what's the point of this insect manifestation of health and safety?

The answer is that nature tends to work in a push-pull fashion. The waggle dance is one way that bees can 'push' each other to venture after certain food sources but, unchecked, this could mean they all end up directing each other into a disaster. So the headbutt-transmitted stop signal serves as a way to rein in the risk. It effectively ensures that the response the colony makes to a threat is proportional to the level of danger it faces.

According to James Nieh, 'This signal is directed at bees who are recruiting for the dangerous food location and decreases their recruitment,' he says. 'Thus, fewer nest-mates go to the dangerous food site. This is important because an individual experiences danger and stops recruiting herself, but the stop signal also enables her to "warn" other recruiting nest-mates too. The end result is that the colony will reduce or cease recruitment to the dangerous food patch in proportion to the threat.'

So bees, it seems, don't just dance for their dinner, they also headbutt their way out of danger!

FACT BOX

What is the waggle dance?

The bee waggle dance was first decoded in the 1960s by the Austrian Nobel prize winning ethologist Karl von Frisch.[*] He christened this unique bee language 'Tanzsprache' and undoubtedly it's one of the wonders of the insect world and what helps to make bees so successful as a social species. In general terms, it works rather like a news grapevine, allowing foragers to share with sister bees information about the direction and distance to food sources or potential new homes.

So what's involved? Rather like 1980s break-dancers on the street, bees perform the waggle dance in front of any other bees who'll stop and watch. Dancing bees move in a figure-of-eight pattern comprising a 'waggle run' followed by a turn to the right to take them back to the starting spot, then another waggle run followed by a turn to the left, again returning

[*] Frisch, Karl von (1967), *The Dance Language and Orientation of Bees*. Cambridge, Mass., The Belknap Press of Harvard University Press.

them to the starting point. Most dances last for 100 such circuits and are accompanied by a waft of chemical signals from the dancer's abdomen to make other workers pay attention.

The direction of the waggle run indicates the direction other bees should fly in, relative to the position of the sun, and the distance is communicated by the duration of the waggle phase of the dance: every 75 milliseconds of dance duration translates into a flight of 100 metres. The bees can also take into account the passage of time and use their body clocks to predict the movement of the sun across the sky to avoid becoming disorientated.

But not all bee observers are quick on the uptake: some are too stubborn to be told, others will cotton on after watching just a few dances, some need a bit more nurturing, while others fly off in a totally different direction despite their education. Pretty similar to humans, really!

Plants don't drink salt water, do they?

They do now

Fertile as the oceans are, if land becomes contaminated with salt it can spell disaster for crops, which shrivel up and die. But fruitful research from Australia now looks set to turn salt toxicity into a marine myth.

Salt is bad for land-based plants because when the plant takes up water from the soil, dissolved salts are drawn in too and make their way up to the leaves. When the water then evaporates from the leaf surface into the surrounding air, the salt is left behind and accumulates. A small amount of salt is tolerable, but if the levels become too high they cause the leaves and shoots to age prematurely. This means that, rather like a loan-shark victim struggling to cover the interest on a mounting debt, the plant ends up investing most of its energy just trying to repair and replace damaged tissue rather than growing effectively.

Now Mark Tester, a plant researcher with Adelaide University,[*] has discovered how to make certain plants salt-tolerant by genetically equipping them with the root equivalent of an inbuilt desalination system. The key to the discovery was a gene called HKT1;1, which makes a protein that can pump sodium, the key component of salt (sodium chloride).

'This gene transports sodium out of cells,' explains Tester. 'So what we've done is to discover a way to increase its expression selectively in the cells that surround the bases of the xylem vessels in the roots. Xylem are the microscopic conduits, like tiny pipes, that carry water and salts from the roots up into the plant shoots. What we have found is that when HKT1;1 activity is increased in the roots of a plant we've been using called *Arabidopsis*, the plants pump more sodium salts out of their xylem and into specialised stores within the root tissue.'

The result is that much less sodium makes its way up the plant, preventing the damage that would normally occur in the shoots and leaves. Consequently, the modified plants will grow

*   *Plant Cell* July 2009: Vol. 21, no. 7, pp. 2163–78 DOI: 10.1105/ tpc.108.064568 happily in highly salt-contaminated soil. So it works in *Arabidopsis*, which is the plant scientist's equivalent of a geneticist's fruit fly and probably tastes equally bad. But what about important food crops? Will they work too?

'Yes!' enthuses Tester. 'We've been able to show this same trick also works in rice, and we're currently testing cereal crops, like wheat, barley and maize.' These latter plant groups have turned out to be slightly trickier to work with because the promoter system – the DNA sequence that the team have used to turn on the sodium-pumping gene in *Arabidopsis* and in rice – does not appear to work the same way in cereals. Instead, Tester and his team have had to find an alternative way to boost HKT1;1 activity, but now they think they are within a xylem-vessel's width of it working.

If they are successful, this discovery will be a major step forward, because currently about one-third of the world's food is grown on irrigated land, one-fifth of which is now significantly affected by salinity problems. This is because 'fresh' water in rivers and streams contains low levels of dissolved salts and minerals and when this is added to a field, the plants use the water (and some evaporates) but the salts are left behind. This causes them to build up over time, eventually making the soil unusable.

As food demands continue to increase, coupled with the effects of climate change such as unreliable rainfall and coastal flooding, the problem is likely to become much worse. So crops that can tolerate conditions like these may be critical in an uncertain future. But are they safe?

'We have checked these plants carefully and there is no evidence that the changes we have made are altering the accumulations of other salts or chemicals within the plant with the exception of a small change to the level of potassium,' says Tester. 'So we're satisfied that these plants do not pose a threat.'

Do woodpeckers suffer brain damage?

Beating your head against a hard surface can be a sign of frustration, yet for a woodpecker it's a fact of life . . .

In the late 1970s, a study carried out by Philip May, Joaquin Fuster, Jochen Haber and Ada Hirschman[*] using high-speed photography (capable of taking 2000 frames a second) revealed that the impact deceleration, when a woodpecker's beak travelling at seven metres per second slams into a tree trunk, can exceed 1000 times the force of gravity (1200 g).

With repeated trauma of this magnitude, it's surprising that the bird's head remains attached to its body, never mind the risk of developing a severe headache, concussion or even brain damage. So why don't they seem to suffer any after effects? Indeed, when other small birds sustain head injuries when they accidentally fly into windows, they usually tumble to the ground and appear to be 'knocked out' for a while before picking

* *Archives of Neurology* June 1979: Vol. 36, no. 6, pp. 370–73; *Lancet* 28 February 1976: Vol. 1, no. 7957, pp. 454–55

themselves up and fluttering off. Why should woodpeckers be any different? The answer is that evolution has equipped them with a number of adaptations that make repeatedly banging your head against a hard surface 20 times per second slightly more tolerable.

Firstly, woodpeckers have relatively small brains which, in contrast to a human, are packed fairly tightly inside their skull cavity. This prevents the excessive movement of the brain inside the skull, which causes so-called 'contre-coup' injuries in humans. These occur when the brain bashes into the skull following a knock on the head. In other words, the head stops, but the brain keeps on moving momentarily afterwards.

Secondly, unlike a human brain, the surface of which is thrown into ridges and folds known as 'gyri' to enable more grey matter to be packed in, the woodpecker's brain has a smooth surface and, through its small size, a high surface area to weight ratio. This means that the impact force is spread over a much larger area, relatively speaking, compared with a human. Again, this minimises the applied trauma. The bird's brain is also bathed in relatively little cerebrospinal fluid, which also helps to reduce the transmission of the shockwaves to the brain surface.

Finally, and possibly most importantly, the woodpecker makes sure that he minimises any side-to-side movement of his head, and this is where May and his colleagues' fast film footage comes in. The team found a tame acorn woodpecker that could be encouraged to perform for their camera if they first bashed out a few words on an old-fashioned typewriter. They watched as the bird first took aim and delivered a number of 'test taps' before unleashing a salvo of strikes, but always in a dead straight line.

This approach is crucial because it avoids placing rotational or sheering stresses on the nerve fibres in the brain. Humans involved in car and motorcycle accidents frequently develop the symptoms of 'diffuse axonal injury' (DAI), where sudden deceleration coupled with rotation literally twists the different parts of the brain off each other like a lid coming off a jar. By hammering in a dead straight line, woody woodpecker avoids giving himself DAI, further minimising the risk of brain damage.

An unresolved issue, however, is that the researchers noted from their photographs that their study subject also took the precaution of closing his eyes just before each strike. But whether this was to keep wood chips out, or the eyeballs in, is anyone's guess!

Cost of keeping me cool? Put it on my bill

In publishing the biological blockbuster that gave us evolution and rewrote the rules of ecology, naturalist Charles Darwin dwelled extensively on birds' beaks. But one of his claims – that toucans are endowed with enormous 10 inchers to attract the opposite sex – turns out to have overlooked a red-blooded effect of a very different kind. Because toucans, scientists now know, are equipped with the world's best built-in biological radiator, which is more efficient at ditching excess heat than even an elephant's ears.

Taking the relative size of a toucan into account, their beaks are up to 40 times larger, in surface area terms, than they should be. In fact, in some specimens, the beak amounts to more than 50% of the surface area of the whole animal. To put that into perspective, a human with similar proportions would have a mouth more than a metre across, much like Mick Jagger.

Although scientists had speculated that this was an adaptation to help the birds peel fruits found in their native Central and South America, or to serve as the avian equivalent of a nuclear deterrent to ward off potential predators, Ontario-based researcher Glenn Tattersall and his colleagues[*] wondered whether this hotly debated oversized appendage might be more to do with heat than fruit, fighting or sex.

Using infrared cameras, the team looked at thermal losses from the birds' beaks over a range of ambient conditions and compared these with the animals' core body temperatures. As the environment became warmer, they found the birds increased blood flow through their beaks, turning them into thermostatic radiators to dump excess heat and keep their body temperatures stable.

Even more cunningly, the amount of bill being used in this way could also be varied. The scientists saw that the beak radiator first turned on when the local temperature exceeded 16 degrees Celsius. At this point, initially just the part of the bill closest to the face warmed up. But as the temperature continued to soar, a progressively larger area of the beak moving away from the face was recruited, until eventually the whole thing

*   Science 24 July 2009: Vol. 325, no. 5939, pp. 468–70
     DOI: 10.1126/science.1175553

was pumping out heat like an infrared beacon.

The researchers suspect that the toucan achieves this by having several sets of incoming blood vessels that can be opened up in turn along the beak. By diverting blood into these vessels and then through the surface tissues of the bill, warm blood is brought close to the body surface, where it releases heat and cools down in the process. Measurements of the rate of heat loss from the birds showed that, depending upon local conditions and wind speeds, between 25% and 400% of the animal's baseline heat production can be lost from the body in this way. In comparison, an elephant's massive ears allow it to dump up to 91% of the heat it produces internally. This suggests that toucans might also use their beaks like this to 'keep their cool' when they are flying, since this consumes large amounts of energy and involves significant muscle activity, which produces 10-to-12 times more heat than when the animal is at rest.

To find out, the researchers also filmed a 10-minute toucan test flight. At the time of take-off the bill temperature was 30 degrees, but within four minutes of becoming airborne this had begun to climb and by 10 minutes it was

37 degrees Celsius. So far from being a sex object, a toucan's bill performs a far more important function – as a hot rod!

FACT BOX

Birds

Apart from big beaks, birds also have a remarkably well developed neighbourhood watch scheme, researchers have discovered recently. Cambridge University zoologists Nick Davies and Justin Welbergen[*] were interested in understanding how reed warblers, who are frequently targeted by cuckoos, learn to fend off the threat. They noticed that younger birds tended to be frightened off by cuckoos, probably because they have evolved to resemble hawks. Older, clearly wiser warblers, on the other hand, are much more audacious and will noisily mob an encroaching cuckoo to stop it laying an egg in their nest.

[*]   *Science* 5 June 2009: Vol. 324, no. 5932, pp. 1318–20
       DOI: 10.1126/science.1172227

The scientists wondered whether the younger birds might be learning to recognise and react to the cuckoo by watching the reactions of their more experienced neighbours. To find out, they placed a series of fake cuckoos at a reed warbler nesting site in the Cambridgeshire fens and also played, through a concealed speaker, the cries made by reed warblers mobbing cuckoos to attract neighbouring animals. Sure enough, the local youth all turned out and perched nearby, watching how the older birds warded off the threat from the fake cuckoo.

Then, to find out how well the locals had learned from the experience, the two researchers subjected birds in neighbouring nests to the same treatment. These animals, which had previously been spectators, reacted far more vigorously to the cuckoo this time, suggesting that they had taken a leaf out of their neighbours' nest and learned to recognise a threat and how to respond to it. 'This explains,' says Nick Davies, 'why there are examples of birds becoming less susceptible to cuckoo parasitism much more quickly than evolution alone would allow. It's achieved through social learning like this.'

Social networking for birds? At this rate they'll all be into Twitter . . .

Sing a song of distance

It's not just humans that like a warm winter holiday, birds do too. And luckily for them – having wings of their own – they're spared the exigencies of economy class travel. Yet although they migrate in their millions to-and-fro between far-flung corners of the globe, scientists really know precious little about this incredible feat of avian navigation. Previous statistics suggested that the average songbird might manage about 150 kilometres a day. Now, thanks to new research, this turns out to be an intellectual flight of fancy: the real rates are almost mythical by comparison.

One of the biggest problems with studying songbird migrations has been how to accurately follow the birds' progress. In the past, tracking equipment was far too bulky for tiny birds to bear without the risk of overloading them, which meant that scientists were forced to rely on marking the animals and then keeping an eye out for their arrival or even trying to pursue them by aeroplane, as one team did. As a result, most attempts to log bird migrations only ended up recording where the animals departed from and when and where they ended up, missing the critical aspect of what happened in between.

But now Bridget Stutchbury from York University in Canada[*] has solved the problem by fitting tiny geolocating bird-backpacks to 14 wood thrushes and 20 purple martins caught in their native Pennsylvania. The backpack devices are about the size and weight of a small coin, contribute less than 5% to the bird's entire weight and record light levels, so the position of a bird wearing one can be accurately pinpointed by referring to the sunrise and sunset times, which vary geographically.

Once equipped, the birds were released and embarked on their winter migrations to South America. The following year, when the birds returned to their US mating sites, some were successfully recaptured, allowing the data to be downloaded from their geolocators and the routes taken by the birds over various time periods to be retraced. 'Never before has anyone been able to track songbirds for their entire migratory trip,' Bridget Stutchbury points out.

* *Science* 13 February 2009: Vol. 323, no. 5916, p. 896
DOI: 10.1126/science.1166664

What was really surprising were the speeds at which the birds were covering the ground. Moving more than 450 kilometres per day, they were travelling three times faster than scientists had previously thought possible. The following spring, the return leg was even more impressive. They got back up to six times faster than during the outward leg, although they did have an agenda: they were heading home to mate and there's a significant competitive advantage to being the first back, because early arrivals have access to the best nest sites and the most food – quite similar to students returning to university at the start of the academic year, you could say.

The researchers were pleasantly surprised by the results. 'We were flabbergasted by the birds' spring return times. To have a bird leave Brazil on April 12 and be home by the end of the month was just astounding. We always assumed they left some time in March,' Stutchbury said.

The researchers also found that prolonged stopovers were common during fall migration. The purple martins, which are members of the swallow family, had a stopover of three-to-four weeks in the Yucatan before continuing on to Brazil. Four wood thrushes spent one-to-two weeks in the south-eastern United States in late October before crossing the Gulf of Mexico, and two other individuals stopped on the Yucatan Peninsula for two-to-four weeks before continuing migration.

There is, however, a sad note to this happy tune. Songbird numbers internationally are in dramatic decline, with some monitoring organisations reporting 70% reductions in the numbers of some species since the 1960s. Human encroachment and destruction of the animals' natural habitats have been blamed as the leading causes, alongside pesticides and poaching, but it's very hard to know how to intervene effectively in the conservation of a species if you don't know why it's threatened in the first place. This is where this study comes in.

'Songbird populations have been declining around the world for 30 or 40 years, so there is a lot of concern about them,' points out Stutchbury. 'Tracking birds to their wintering areas is also essential for predicting the impact of tropical habitat loss and climate change. Until now, our hands have been tied in many ways, because we didn't know where the birds were going. They would just disappear and then come back in the spring. It's wonderful to now have a window into their journey.'

Another important question that needs answering is what do birds do about bad weather? The next step Stutchbury and her colleagues are planning is to marry up the songbird migration routes with weather reports to see how individual birds respond when faced with a storm. Will it ruffle their feathers and leave them in a flap, or will they rise above it?

Magnetism's invisible, isn't it?

More than 50 different animal species, including mammals, turtles, lobsters, fish, birds and bees are known to possess a built-in compass of some sort that allows them to plug into the earth's magnetic field so they can find their way around. Salmon, for instance, appear to use this trick to return from the ocean to reproduce in the rivers in which they were born, homing pigeons and bats both go off course if exposed to strong magnetic fields, as do bees and ants, and some bacteria even seem to know their north from south.

But despite many years of study, no one has yet discovered exactly how any of these species do this. One theory is that magnetically sensitive deposits of iron-containing compounds – like magnetite – are used, and some animals do show accumulations of iron particles in some parts of their bodies. More controversially, other scientists have suggested that some species may actually be able to see magnetic fields, although, attractive as this sounds, it's remained very much a magnetic myth because scientists haven't managed to find any sort of chemical reaction that would be sufficiently sensitive to a relatively weak field like the earth's.

That is, until now, because Oxford scientist Peter Hore[*] has recently uncovered a quantum mechanical chemical trick that some animals could be using to make magnetism visible. He proved this using a large, purpose-built molecule called CPF. At one end, this has a chemical group called a carotenoid, which consists of a long chain of carbon atoms flanked by carbon atom rings. This is linked to a large molecule called a porphyrin ring, which sits at the centre of the structure. Bonded onto the other side of this is a carbon 'football' known as a fullerene. Seen from the side, it looks a bit like a miniature molecular weather vane, although with the important distinction that it's sensitive to magnetic fields rather than wind direction.

Molecules of CPF were suspended in liquid crystal and frozen at minus 80 degrees Celsius to lock them into position. They were then excited using light of a certain wavelength while tiny magnetic fields, on par with those produced by

[*]   *Nature* 15 May 2008: Vol. 453, pp. 387–90 DOI: 10.1038/nature06834

the planet, were applied in different orientations. Incredibly, the molecules altered their chemical behaviour according to the direction of the magnetic field.

There were several stages to the process. First, the CPF molecule was excited by the light. This made the chemical groups at each end of the molecule (the carotenoid and the fullerene) temporarily become free radicals, which means that they each contain an unpaired electron. These electron radicals are spinning, and normally they spin in opposite directions to each other. But because the CPF molecule is quite long (as molecules go), these free radicals are separated by a distance of about 3.5 nanometres (3.5 billionths of a metre), which means that their spins can be affected by a magnetic field applied in the right direction. When this happens, one of the electrons can flip over, so the two end up spinning in parallel, rather than in opposite directions.

This quantum flip effectively unbalances the molecule, making it adopt a form called a triplet state. In this condition, before it can release the extra energy it absorbed and return to its starting state, it needs to go through an additional reaction step, which means it remains in this altered state for slightly longer, giving the chemists an opportunity to detect it. In the eye, if a similar chemical process exists, this prolonged altered state could be picked up by some kind of signalling molecule and then relayed to the brain to make the animal aware of the orientation of the surrounding magnetic field.

Of course, the present experiments are a highly artificial situation, not least because eyes don't contain CPF. But birds' eyes do contain light-sensitive chemicals called cryptochromes, which could behave in the same way, adopting an altered 'radical state' when the animal is aligned with the magnetic field. 'This could trigger a chemical to change shape, which could in turn kick-start other biochemical processes to enable a bird to see the earth's magnetic field,' says Hore. So maybe there is something to saying that certain people have a magnetic personality after all . . .

FACT BOX

Magnetism and animals

Although scientists still don't understand how animals detect the planet's magnetic field, evidence of them doing so is everywhere to see, including, as German and Czech researchers have recently shown, on the nearest cattle farm, and even on Google Earth.

In 2008, Sabine Begall, from the University of Duisberg-Essen,[*] used Google's satellite profile of the planet to find pictures of cattle grazing in fields at 308 locations on six continents: Europe, Africa, India, North and South America and Australia. These images, containing over 8500 cud-chewing cows, were analysed to see in which direction the animals were pointing in their fields. At the same time, field trips were made to observe more than 2900 deer in the Czech Republic.

Astonishingly, both the cows and the deer were arranging their bodies to point along the line of magnetic north–south. The effect

[*]   *PNAS* 9 September 2008: Vol. 105, no. 36, pp. 13451–55
       DOI: 10.1073/pnas.0803650105

was slightly less pronounced over Africa, but this would be expected of a magnetic effect because the earth's field is slightly weaker there. Amongst the deer in the Czech Republic, the majority of the animals were found to stand (and sleep) pointing north. About a third of them arranged themselves in the reverse (south–north) direction at any one time, which might be a strategy to help them to look out for predators approaching from behind.

These findings were highly statistically significant, suggesting that the effect is real. Begall scrutinised the images looking for other reasons to explain why the animals should all line up like this, but found none. She was able to dismiss prevailing winds on the grounds that animals all over the world were doing the same thing simultaneously, yet the prevailing winds vary in direction depending upon location. Similarly, she wondered whether the animals were adjusting their sun exposure as a form of temperature control or whether they were positioning themselves to avoid being dazzled by bright light, but again, neither of these possibilities could explain the planet-wide observation, which means only magnetism remains as an explanation.

For the moment, science is left with an interesting enigma. Why should these large, social ruminants, all with extensive ranges, feel compelled – or should that be repelled – to align themselves with the earth's magnetic field? Perhaps this is the bovine or cervine equivalent of geophysics – a way to methodically plod around a field to find food most efficiently by working in rows? Or maybe the magnetic field alters vision, making predators easier to spot? In humans, scientists have reported that the EEG (brainwave) pattern is slightly different depending upon whether someone sits facing north–south or east–west, and the rapid eye movement (REM) phase of sleep also changes subtly according to whether sleepers are slumbering in an east–west or north–south repose. But what the significance of this is, or even how it happens, no one knows. Observation and explanation, you could say, appear to be poles apart . . .

It's not what you say, it's how you say it

In 1968, the Bee Gees penned a set of lyrics which included the lines 'It's only words, and words are all I have . . .' The result was a hit for them and also for the parade of artists who have since covered and re-covered that very song, 'Words'. At the same time, many of us would probably agree that words, used the right way, can be pretty powerful.

As it turns out, the Bee Gees actually missed a trick, because new research has revealed that there's far more to a touching oratory than just the sounds we hear. In fact, a study from Canada shows that we genuinely do feel 'moved' by speech, because the brain picks up the sensations of air rushing past the body and uses this to bolster its understanding of what's being said.

For English speakers, this mostly applies to sounds like 'pa' and 'ta', produced by placing the tongue towards the front of the mouth. Critically, these sounds are also accompanied by a puff of air as you make them, noticeable if a hand is held in front of the lips as the letters are articulated. In isolation they're quite easy to make out, but throw in a cocktail party or a live band in the background and these 'pa' and 'ta' noises can be quite hard to discriminate from the similar sounds 'da' and 'ba' – except that these latter sounds aren't associated with any significant rush of air when you say them.

This got Bryan Gick and Donald Derrick, two researchers who are based at the University of British Columbia,* wondering whether the brain might make use of these effects as the aerodynamic equivalents of lip-reading. To find out, they rigged up a sound experiment in which 66 blindfolded volunteers listened through headphones to recordings of a male voice saying 'pa', 'ta', 'da' and 'ba' sounds. The subjects were asked to indicate, by pressing buttons, which sounds they had heard. At the same time as the sounds were being played, on some occasions the researchers also squirted brief pulses of air from a tube onto one of the subjects' hands or their necks.

*   *Nature* 26 November 2009: Vol. 462, pp. 502–04
DOI: 10.1038/nature08572

When presented alongside 'pa' and 'ta' sounds, which are naturally accompanied by an air-rush, the puffs of air from the tube significantly boosted the accuracy of the subjects' hearing. They went from being right about 65% of the time to being right more than 75% of the time.

That was when the air puffs accompanied the sounds they should have done. But what would happen if a subject experienced an air puff when they heard a 'ba' or 'da' sound that would not normally be associated with any such gaseous emission? Incredibly, by playing this trick, the researchers found they could fool the brains of the subjects into hearing the wrong thing. Their accuracy fell from 85% in the absence of any air puffs to about 70%. Another surprise was that it didn't seem to matter which part of the body the air puff was hitting – hand or neck, both had the same effect. As Bryan Gick puts it, 'It sort of challenges this traditional idea that you see with your eyes and that gets processed by a particular part of your brain and you hear with your ears and that gets processed by another part of your brain. It looks like our brains just take everything in.'

In other words, it seems that in making sense of what's being said, the brain takes its lead partly from the Bee Gees and also injects a dose of a-ha's 'Touch Me'. It certainly adds a whole new dimension to the phrase 'breathing down someone's neck' . . .

Babies learn to talk like mum, in utero

Anyone who's ever spent time with a toddler knows only too well how many words they've already mastered. New linguistic gymnastics appear on an almost daily basis, most of it mimicked from mum and dad. Indeed, by the time the average English speaker reaches the age of 17, they probably know more than 60,000 words. But what about their accent?

The way we sound singles us out almost as distinctively and uniquely as the way we look, but where did we learn to speak like that and why? At the simplest level, accents are just the way we choose to make the sounds that others

Does this nappy make me look fat?

comprehend as language. And because humans are social creatures that strive to bond and to imitate one another, which is – after all – how we learn, scientists had thought that accents were something we acquired as language developed.

But new research shows that this is a tongue-twistingly massive myth, because – *Ooh, la la* – scientists in France and Germany have now found that babies are born with their accents already in place. Kathleen Wermke, a behavioural scientist at the University of Würzburg,[*] made the discovery when she and her colleagues recorded the cries of 60 French and German newborns, all of whom were under three days old. They fed the coos and yelps into a computer program that analysed the frequency and melody contours of the cries.

In each case, a characteristic 'signature spectrum' emerged, which strongly correlated with the nationality of the newborn. Those with French mothers produced a rising melody contour, while German babies had a falling melody contour. In other words, the cries either increased or decreased in pitch with time. Astonishingly, these contours matched the sonic signatures of the parents' native languages.

[*]   *Current Biology* 5 November 2009: Vol. 19, no. 23, pp. 1994–97
      DOI: 10.1016/j.cub.2009.09.064

How were the babies picking up their accent in the first place? The simple answer is by eavesdropping on mum's conversations. We know that sounds made by a pregnant mother, as well as speech and other noises originating from outside the body, can make it into the uterus to be picked up by a baby. This was confirmed in 2004, with the help of a pregnant sheep, by two researchers at the University of Florida, Ken Gerhardt and Robert Abrams.[*]

They placed a tiny microphone inside the ear canal of a developing lamb. Then, 64 spoken sentences were played on a loudspeaker placed in the open air next to the pregnant sheep while a recording was made simultaneously from the microphone in the lamb's ear. When the lamb-ear recordings were then played back to a group of 30 human volunteers, more than 30% of what was said could be understood. Critically, however, the recordings revealed that the low-frequency sounds were picked up best. According to Gerhardt, in the context of what might be transmitted to a developing human baby, 'That means they are more sensitive to the melodic parts of speech than

[*] *Audiology & Neurotology* November–December 2003: Vol. 8, no. 6, pp. 347–53 DOI: 10.1159/000073519

to pitch.' So far, so good. But does a developing human baby really 'hear' *in utero*?

'Yes,' say scientists, who think that the hearing system is probably wired up and working by 30 weeks gestation. Barbara Kisilevsky, from Queen's University in Canada,[*] played sounds to 143 foetuses aged between 23 and 34 weeks of gestation while watching their reactions *in utero* using ultrasound. Babies aged over 30 weeks, she found reacted to the sounds, but the less-developed individuals didn't.

It seems, therefore, that developing babies can hear what mum's nattering about. But why bother copying her? Subversion on the part of the baby, suggests Kathleen Wermke. To make sure its mother loves it. 'Newborns are probably highly motivated to imitate their mother's behaviour,' she says, 'in order to attract her and hence to foster bonding.'

So why do it in cries? you ask. 'Because,' she points out, 'melody contour may be the only aspect of their mother's speech that newborns are able to imitate; this may explain why we found melody contour imitation at that early age.' So

[*]   *Early Human Development* June 2000: Vol. 58, no. 3, pp. 179–95
    DOI: 10.1016/S0378-3782(00)00075-X

watch out: not only is your future son or daughter listening intently to everything you say towards the end of pregnancy, they're also learning to take you off!

FACT BOX

Nature vs. nurture?

Apart from accents, there are many other things that babies pick up from their parents, or appear to. But a big problem in biology is how to disentangle the effects of inherited genes – in other words, *nature* – from the effects of the local environment in which the child grows up – that is, *nurture*.

For instance, scientists have noticed a link between maternal smoking and babies born with a low birthweight, and there is also an increased likelihood that a smoker's child will display antisocial behaviour when it grows up. But to what extent might genes be responsible for both of these effects and, conversely, how much of a role does the environment play?

Mothers who smoke when they are pregnant might be more likely to have behavioural problems themselves or a genetic tendency to deliver low birthweight babies that also makes them smoke.

Anita Thapar, a researcher at Cardiff University in Wales,[*] recently came up with an ingenious way to solve this problem. She compared maternal smoking with the birthweights and subsequent antisocial behaviour rates amongst 779 children. But these weren't any old children: these study subjects were all conceived by *in-vitro* fertilisation (IVF) and over 200 of the babies were born from *donor* eggs. In other words, the babies born from these eggs were genetically unrelated to their 'mothers' but still developed exposed to the environment inside her. This neat study design meant that, in assessing the outcomes for the children, genetic factors – *nature* – could be effectively separated from *nurture* effects – maternal and environmental factors.

What emerged from the analysis was a very

*   *PNAS* 17 February 2009: Vol. 106, no. 7, pp. 2464 -67
DOI: 10.1073/pnas.0808798106

strong relationship between maternal smoking and low birthweight in both children who were genetically related to their mothers and the genetically unrelated children born from donor eggs. This proves that the effect of smoking on birthweight is not down to genetics, but must be due to the toxic effect of smoking itself.

Antisocial behaviour, on the other hand, was only associated with maternal smoking in children who were genetically related to their mothers; there was no statistical association amongst the children born using donor eggs. In other words, a genetic tendency present in the child and passed down from the mother is triggering the antisocial behaviour. This same genetic tendency may also explain why mum smoked while she was pregnant in the first place.

'This shows the importance of inherited factors in the association between prenatal smoking and offspring behaviour,' says Thapar. 'This suggests that gene-environment correlation is important in explaining this association.'

Blind as a bat? Or should that be rat?

Being called 'blind as a bat' is usually a term reserved for people who could miss an elephant at five feet, trip over their own toes or even fail to notice their own spectacles sitting on their nose. But it turns out that this is a rather short-sighted insult, because new research reveals that some bat species actually have quite good eyesight.

Bats come in two types: *little*, known as the Microchiroptera, which hunt at night using echolocation and are insectivorous, and *large*, the Megachiroptera or fruit bats (flying foxes), which also forage at night but are sometimes active during the day and at twilight. Traditional bat wisdom, going back over 100 years and based on microscope studies, says that – unlike humans, who have both rods and cones in the retina to enable us to see reasonably in the dark (with the rods) and in colour during the day (with the cones) – both of these bat groups have a retina containing only rods. And since rods produce generally poor visual acuity, and they can't discriminate between different colours, bats have repeatedly got the gong for being the optically challenged members of the animal world and an ophthalmic insult coined in their honour.

For the night-active and insectivorous Microchiroptera bats, this is probably not an unreasonable conclusion, although having poor vision isn't really a problem when you're blessed with a built-in sonar most submariners would kill to get their hands on. But flying foxes don't use echolocation like this, and they aren't always only active at night either. Researchers have reported that, during the day, they keep a constant eye out for predators, groom and interact with one another, occasionally rearrange their roosting sites to stay out of the sun and, from time to time, younger individuals use the daytime for flight training. Meanwhile, other studies have shown that these animals devote a large region of their brains to visual processing and have forward-facing eyes that fixate on objects of interest.

None of these behaviours seem to square with an animal that is supposed to have limited night-sight and be virtually blind in daylight, which is what prompted German researcher Brigitte Muller and two colleagues* to blink and then

* Brain, Behavior and Evolution August 2007: Vol. 70, no. 2, pp. 90–104 DOI: 10.1159/000102971

take a closer look at what was going on in these animals' eyes. They examined specimens from six different species of fruit bats that had been killed for other purposes. To see what was going on in the retina, they used colour-coded antibodies that could distinguish the different chemicals, called opsins, which are contained in rods and cones where they help to turn lightwaves into brainwaves.

Predictably, lots of the cells picked up by the antibodies were long thin rods. In some cases there were nearly one million of these per square millimetre of retinal tissue. But amongst these were cells lighting up that were clearly cones. They were rare, comprising only about one in every 200 rods (0.5%), but sufficiently numerous to mean that these bats can potentially see better than most rats.

The cones the scientists saw were actually made up of two different populations. There were L-cones, which see green, and in three of the bat species, an additional rarer type of cone called an S-cone, which can see blue light. This means that some of the fruit bats studied are neither blind nor colourblind. So next time you need to insult someone's eyesight, consider 'blind as a rat' as a more accurate alternative to mocking the humble bat's optic abilities!

FACT BOX

How nocturnal eyes use DNA to make light work of seeing in the dark

Mammals like bats, cats and bushbabies, which are all active at night, face a significant challenge when it comes to making their eyes work, because compared with day-active animals, nocturnal species generally have less than one-millionth of the amount of light to see by. But recently, scientists have shown that these animals solve the problem in a very unusual way: by using the optical properties of their own DNA.

A major anatomical issue that affects the eyes of all of us is that our retina is effectively 'inside out'. In other words, the energy-hungry rod and cone cells that detect light are positioned closest to the blood vessels at the back of the eye so that they can pick up sufficient oxygen to meet their needs. But this means that, to reach them, light first has to filter through several layers of supporting cells that are positioned closer to the front of the eye and whose job it is to transmit the information from the rods and cones back to the brain. And because nocturnal animals are working under such low light levels, this can seriously hamper vision.

So what's the solution? By examining rod cells from nocturnal animals under the microscope, Boris Joffe, from the Ludwig-Maximilians University Munich,[*] noticed that it's not just the retina that is inside out – the DNA inside these cells is too.

With a few exceptions, every mammalian cell contains a complete copy of the animal's genome, all packed into about two metres of DNA which is itself tightly coiled inside the nucleus, a button-shaped structure at the

*   *Cell* 17 April 2009: Vol. 137, no. 2, pp. 356–68
      DOI: 10.1016/j.cell.2009.01.052

heart of the cell. Normally, this nuclear DNA is organised such that the parts of the genome being used by the cell are clustered centrally in the nucleus, and the inactive material (including the so-called junk or non-coding DNA) is kept in a tightly wound form, called heterochromatin, around the outside. But in the rod cells of night-active animals, this situation is reversed. The dense, inactive heterochromatin is in the middle of the nucleus while the active genetic material is arranged around the outside. Daytime-active animals like humans, on the other hand, lack this specialisation.

The big question, of course, is why? It turns out that tightly coiled heterochromatin DNA has a much higher refractive index, meaning that it can focus light more powerfully than the more open structure of the active DNA. This means that, in nocturnal animals, the centres of the rod cells behave like miniature lenses, cutting down scattering and helping to funnel photons (light) into the light-sensitive tips of the rod cells. Without this effect, computer models show, the light paths diverge across much larger areas of the retina, reducing the likelihood of detection and cutting sensitivity.

Elegant as it is, when first proposed the idea proved so radical that some members of the scientific community took quite a bit of convincing. 'People laughed at first,' says Joffe. But the argument does make compelling evolutionary sense. All nocturnal mammals have this DNA adaptation in their eyes, which means that it must have arisen early on during mammalian evolution, probably in the region of 100 million years ago.

At this time, mammals were just beginning to appear and most likely adopted a nocturnal lifestyle to enable them to escape the clutches of carnivorous reptiles which, being cold-blooded and reliant on heat from the sun, would have been less active at night. But diurnal (day-active) animals, like humans, have lost the trait again during our evolution because it confers no benefit under the high-light conditions for which our sight is best adapted.

You never make the same mistake twice, do you?

James Joyce once famously described mistakes as 'portals of discovery', but now scientists have discovered that when it comes to tip-of-the-tongue moments, the claim that you don't make the same mistake twice is a myth. In fact, in this situation, one mistake begets another.

Most people agree that there are very few things more frustrating than the excruciating sensation of knowing you know a word for something but can't for the life of you remember what it actually is. It's like having an intellectual itch you can't scratch and it continues until either you manage to remember the word you wanted, or someone puts you out of your misery and tells you. The sense of relief you experience when they do is so totally overwhelming that you honestly can't believe that you'll ever forget that word again. But then you do, the very next time you want to use it!

Don't believe me? Well, anecdotally, that's what

people – and scientists – are saying (assuming they can find the right words, of course). For many years, scientists had thought that the act of successfully surmounting a tip-of-the-tongue moment should ensure that the lost word would be better remembered in future. Indeed, the American psychologist Edward Thorndike wrote in 1913, 'When a modifiable connection between a situation and a response is made and is accompanied or followed by a satisfying state of affairs, that connection's strength is increased. When made and accompanied or followed by an annoying state of affairs, its strength is diminished . . .'

So why do tip-of-the-tongue experiences just keep happening, then? According to Amy Warriner and Karin Humphreys, two researchers at Canada's McMaster University,* it's because rather than learning *from* our mistakes, in this situation we actually end up learning how better to *make* the mistake itself!

The research duo reached this rather unnerving conclusion when they subjected 30 student volunteers to a barrage of 1500 potential tip-of-the-tongue opportunities. The students were presented

*   *Quarterly Journal of Experimental Psychology* April 2008: Vol. 61, no. 4, pp. 535–42 DOI: 10.1080/17470210701728867

with statements on a computer screen provoking them to recall certain words. These statements were carefully crafted to elicit tip-of-the-tongue experiences, which is best done by focusing on seldom-used words or expressions. As a control, some fake words were included to make sure the volunteers were playing by the rules.

After reading each statement, the participants were told to press one of three buttons: 'Know', if they knew the word the scientists were after, 'Don't Know', if they didn't, or 'TOT' (tip-of-the-tongue) if they knew they knew the word but couldn't articulate it. Whenever they experienced a TOT, the team then let them suffer for either 10 seconds or 30 seconds before telling them the word they were looking for.

Two days later, the students repeated the test using exactly the same statements. If Edward Thorndike was right, the words that had tongue-tied the participants first time round ought to be a doddle a day or so after. But in reality, many of the volunteers experienced TOTs again on the same statements that had tripped them up two days before. Even more shocking was that in cases where previously a student had been made to wait 30 seconds before being told the word they were looking for, they were 50% more likely to experience a repeat TOT than with words for which they had been made to wait only 10 seconds.

The researchers think that this is because the intellectual exercise associated with trying to retrieve the TOT word actually makes the brain learn to forget it. In effect, the nerve connections that would normally lead to the retrieval of the correct word have been misrouted. And so, by repeatedly activating this aberrant neurological pathway, the brain is making it stronger and therefore more likely to cause the same problem again. 'Music teachers know this principle,' says Humphreys. 'They tell you to practise slowly. If you practise fast, you'll just practise your mistakes.'

So what should you do to deal with an aggravating TOT? Don't keep racking your brains for the right answer, say the researchers. Quickly ask a colleague for the missing word, or look it up on the internet. Then repeat it to yourself, out loud or in your head. But don't dwell on it if the answer's not forthcoming. Otherwise, Humphreys points out, 'You'll keep digging yourself the wrong pathway.'

FACT BOX

Tasty TOT titbit

Tip-of-the-tongue experiences have also been researched recently in volunteers with a rare form of a condition called synaesthesia which causes them to experience specific tastes in their mouths when they are about to say certain words (and not just the names of foods either!). Characteristically, the same words always trigger the same flavour sensations.

This is known as gustatory synaesthesia and scientists think that, in people who have it, there are extra connections between the brain area that processes speech and the brain region that decodes flavours. Consequently, when a word is selected for speech it also causes the brain to fire off a flavour sensation. For these people, conversation can feel like quite a mouthful . . .

But two UK researchers, Julia Simner from Edinburgh University and UCL scientist Jamie Ward,[*] wondered what would happen to these

*   *Nature* 23 November 2006: Vol. 444, p. 438
DOI: 10.1038/444438a

individuals – who are otherwise neurologically normal – if they were to have a tip-of-the-tongue experience? They asked six gustatory synaesthetes to look at a series of pictures of rare or unusual objects designed to elicit tip-of-the-tongue moments. Intriguingly, whenever this happened in the subjects, although they couldn't say the word they were looking for, they could still 'taste' the word they wanted. When the study was repeated on the same patients using the same images a year later, many of the subjects had TOTs on the same words as they had before – and reported the same flavour sensation.

This shows that, whatever the extra connections are in the brains of people with this form of synaesthesia, they must lie downstream of the system that selects a word to use but upstream of the system that translates this thought into a physical word we can say. How that's for research you can get your teeth into?

Your thoughts are no longer your own

Most people are comfortable with the idea that what goes on in your own head is for your mind's eyes only. That's because, put simply, thoughts are nothing more than clusters of nerve cells firing off in a certain sequence. And, unless you choose to tell someone what you're thinking, this means it's impossible for anyone else to know. At least, that *was* the case. Because recent advances in brain scanning and mapping techniques mean that scientists are moving ever closer to being able to decode what's going through the heads of the average person, ushering in an era when your thoughts will no longer be your own.

This was elegantly demonstrated by a team of US scientists led by University of California Berkeley researcher Jack Gallant.[*] He and his colleagues have developed computer software that, based on the pattern of brain activity in a human subject, can predict with 90% accuracy what image – from a selection of thousands – a person is looking at.

[*] *Nature* 20 March 2008: Vol. 452, pp. 352–55 DOI: 10.1038/nature06713

The technique uses functional magnetic resonance imaging (fMRI), an existing technology that can decode nerve activity by comparing how much energy is being consumed by each region of the brain at any one time. When a part of the brain becomes busily engaged processing a piece of information, its oxygen demand shoots up, which boosts the local blood flow, and this is what the scanner can see.

What the Berkeley team have done is develop a system that looks very closely at the regions of the brain concerned with visual processing. By showing human subjects a series of 1750 different photographs and then comparing how the brain responded to each image, the team were able to match up, at very high resolution, how different aspects of the photographs produced correspondingly different brain activities. After five hours, the system had learned how the brains of the human volunteers were responding to individual details of the visual images.

Next, to find out how well it really knew its subjects, the researchers put the software to the test. They asked it to predict the pattern of brain activity that would be produced when the subjects viewed 120 different images that they had never seen before. The predictions were then compared with the scan results collected when the subjects viewed the images for real. Astonishingly, the software was getting it right 92% of the time in one volunteer and 72% of the time in the other. This is compared with a success rate of 0.8% (1:120) which would be achieved if random guesswork were being used.

To find out whether the system could cope with much larger numbers of images, the team then tested its performance against a suite of 1000 photographs, achieving a similar success rate of 82%. They also re-tested the same participants two months and then 12 months later and found that the system was still working as well as it had previously, indicating that once it had learned how a subject's brain processes visual information, the system could continue to work reliably into the future.

Thinking ahead, Jack Gallant points out that these results show 'that it may soon be possible to reconstruct a picture of a person's visual experience from measurements of brain activity alone.' In other words, go one step further and, by reading off the pattern of brain activity, translate what is in someone's mind's eye into a physical picture on a computer screen. 'Imagine a general brain-reading device that could reconstruct a picture of a person's visual experience at any moment in time.'

Which means that 'turning your dreams into reality' could take on an entirely new meaning in future.

'Brain' box

The human brain contains about 100 billion nerve cells, each making an average of 1000 connections to other brain cells. These cellular links are known as synapses and when a brain cell becomes active, the synapses squirt out nerve transmitter chemicals (neurotransmitters) that lock onto adjacent nerve cells and alter their activity. By making, breaking or altering the relative strengths of these connections – a process called long-term potentiation (LTP) – the brain creates mouldable neural networks that can store and process information, including making memories.

This means that when it encounters certain stimuli, the brain responds with characteristic patterns of activity, which is what Jack Gallant and his team were seeing in their fMRI study.

Brain dead . . . or maybe not?

When someone's said to be brain dead, doctors usually mean a patient who is in what's called a persistent vegetative state. This is where an individual can appear to be awake, their eyes can be open, they even periodically seem to go to sleep, but they remain completely unaware, unreactive and unresponsive to the world around them. Most remain in this state permanently.

People who fit this category are usually victims of severe head injuries sustained in road traffic accidents or falls, or patients who have suffered a lack of oxygen supply to the brain through drowning, carbon monoxide poisoning, heart attacks, or strokes. Under these circumstances, there is usually significant damage to the brain. The person remains physically 'alive' because the primitive parts of the nervous system, which control heart rate, blood pressure and breathing, are still working. But the destruction of the 'higher' brain regions that make a healthy individual conscious renders the person permanently vegetative: awake, but unaware.

Or so we thought. Because a Cambridge-based brain researcher, Adrian Owen,[*] recently succeeded in communicating, using a brain scanner, with a patient written off as brain dead more than five years previously. The team had wondered whether some vegetative patients are unresponsive because the regions of the brain required to initiate bodily reactions have been too badly damaged. They reasoned that if someone was in this position they might nonetheless be able to see things in their mind's eye or imagine themselves performing certain tasks if instructed to do so.

So over a three-year period, the researchers screened 23 patients diagnosed previously as being in vegetative states. With the participants in a brain scanner, they asked them to perform two imaginary tasks. The first was to see themselves playing tennis. This triggers high levels of activity in a region of the brain called the supplementary motor area (SMA), which is concerned with planning movements. The second task was to imagine walking around home or a familiar place. This activates a different brain region – the

*   *New England Journal of Medicine* 18 February 2010: Vol. 362, pp. 579–89 DOI: 10.1056/NEJMoa0905370

parahippocampal gyrus – which is concerned with navigation and sense of direction.

The researchers had already demonstrated, using healthy volunteers, that a functional MRI (fMRI) scanner could detect the differences in brain activity between these two tasks. This meant that if any of the vegetative patients could understand the instructions, such as 'start imagining playing tennis now', then they ought to be able to alter their thoughts accordingly, which the scanner would be able to pick up.

When the results from the patients began to roll in, the researchers were shocked. Four of the 23 showed signs of awareness and, incredibly, by thinking about tennis to indicate a positive and walking around his house to indicate a negative, one of the patients was even able to answer 'yes' and 'no' questions such as, 'Is your father's name Thomas?'

According to Adrian Owen, 'We were astonished when we saw the results of the patient's scan and that he was able to correctly answer the questions that were asked by simply changing his thoughts. Not only did these scans tell us that the patient was not in a vegetative state but, more importantly, for the first time in five years it provided the patient with a way of communicating his thoughts to the outside world.'

What this study shows is that not all patients who appear to be brain dead actually are. But thanks to an imagined game of tennis or a trot around home it might be possible to reach those that are aware and reopen a window into the world for them.

Face it! Emotions aren't always easy to read

Ask a friend to look at a photo of a face and you'd expect them to be able to tell you what sort of mood the pictured person was in, wouldn't you? Charles Darwin felt the same way and said so in his 1872 book *The Expression of the Emotions in Man and Animals*. Facial expressions, he thought, are a universal window into emotion, unaffected by creed or culture.

Scientists have since taken this belief much further, leading to the creation in the 1970s of a concept called FACS (facial action coding system). Using this approach, a set of universal facial images were created to represent the seven basic human displays of emotion: 'happy', 'sad', 'angry', 'fearful', 'neutral', 'surprised' and 'disgusted'. These, they argued, should be recognisable by everyone. But some scientists frowned on this convenient claim of a cross-cultural appreciation for facial expression, so University of Glasgow psychologist Rachael

Jack decided to put it to the test.* And, unfortunately for Darwin, it turns out that some of these so-called 'universal' facial expressions are commonly confused by different cultures. Or even lost in translation, you might say.

The scientists flushed out this myth by recruiting a group of western Caucasian volunteers and a similar-sized group of East Asians, most of them Chinese students recently arrived in Britain to study. The subjects were asked to look through a collection of 56 images displaying the seven key FACS-coded facial emotions. The picture mix included both oriental and western faces.

As they regarded each image, the participants were asked to decide what emotion was being displayed. While they were doing this, the researchers also monitored the subjects' eye movements to discover the order, timing and priority with which they scrutinised the different regions of the faces presented. The western Caucasian participants performed with consistent accuracy across the entire panel of faces and emotions, but the East Asian subjects fared much less well, confusing 'disgusted' with 'angry' faces,

*   *Current Biology* 13 August 2009: Vol. 19, no. 18, pp. 1543–48
    DOI: 10.1016/j.cub.2009.07.051

Stripping Down Science

and substituting 'surprised' for 'fearful' faces.

But why? The reason became apparent when the team graphically reconstructed the eye movements made by both groups when they looked at the images. It was clear that while the western Caucasians tended to survey the whole face – and apply roughly equal weight to most regions – the East Asians strongly biased their observations in favour of the eye region of each face. As a result, the Asian participants had a harder time distinguishing emotions in which the eyes look similar but the set of the mouth differs. For instance, faces showing fear or surprise both have big, wide open eyes, but the mouths are very different. 'It would be difficult to distinguish between the two . . . if you didn't look at the mouth,' says Rachael Jack.

More intriguing is the fact that the evidence for this dramatic finding has already been staring us all in the face on the internet for many years in the form of the 'emoticons' users add to the things they write online to convey their emotions. Westerners tend to employ symbols that use the mouth to convey emotional states such as :) to imply happiness, or :(for sad. But look at the blog post of an easterner and you'll find the eyes doing the work instead, as in ^.^ for happy and ;.; for sad.

The big question, though, is how this difference has arisen in the first place and why, which is what the researchers are now investigating. Convinced that the effect is cultural rather than inherited, they are mapping out how emotions are manifested in Asian faces, as well as trying to find out what happens to children born in the west to Chinese parents. Can they switch expression-reading strategies to suit their immediate surroundings?

If the cause is a cultural one, a likely explanation for the observed east–west divide is that some Asian cultures consider overt displays of emotions to be impolite. For this reason, these people may have learned to focus on more subtle cues arising from around the eyes rather than relying on mouth movements to do the talking.

Pollution superhighway skyward

In attempting to understand how the atmosphere circulates around the earth, and particularly how pollutants are carried aloft, researchers recently realised that there was a monsoon-sized hole in their atmospheric argument.

Scientists divide up the earth's atmosphere into a number of different layers. Between the ground and a height of up to 15 kilometres is the so-called troposphere, which is thickest at the equator and thinnest at the poles. It's created by heat from the earth's surface causing rising currents of warm air that trigger winds, rain and other weather phenomena. Above the troposphere is a layer called the stratosphere, which extends up to a height of about 51 kilometres above the planet's surface and also contains the ozone layer, which shields us from the sun's ultraviolet rays. Between the troposphere and stratosphere is a layer called the tropopause, which is the point at which air from the planet's surface stops rising and becomes almost completely dry.

Researchers have found that the air in the troposphere and the air in the stratosphere largely remain separated from each other and only able to mix in appreciable amounts over the tropics. Canadian climatologist Alan Brewer and Oxford physicist Gordon Dobson first identified this phenomenon over 60 years ago and it is now known as the Brewer-Dobson circulation in their honour. Consequently, scientists thought that pollutants present in the troposphere (arising from the earth's surface) would only be able to enter the stratosphere if they first made their way to the tropics in order to join air rising in the Brewer-Dobson circulation. However, en route to the tropics, many pollutants are soaked up by the sea long before they can be carried aloft, which helps to keep the contamination out of the stratosphere.

But now scientists have discovered that the Asian monsoon – an annual storm system that occurs when warm, moist air from the Indian and Pacific Oceans is drawn in over Asia by the temperature difference between the land and the sea – might be providing an alternative portal to the upper atmosphere, ferrying with it water, carbon particles, oxides of sulphur and nitrogen and also ozone-damaging halogen compounds produced by Asia's burgeoning heavy industries.

This atmospheric body blow was revealed by US researcher William Randel, based in Boulder, Colorado.[*] He used a satellite-based system called ACE-FTS (atmospheric chemistry experiment Fourier transform spectrophotometer) to track the levels of the gas hydrogen cyanide around the globe. Scary as it sounds, hydrogen cyanide is produced quite normally when plant-based material is burned, for instance when forests are cleared for agriculture. Usually, it is quickly mopped up by the ocean, so the levels are kept low. But the satellite images told a different story: spreading out across the stratosphere above the top of the monsoon was a very high concentration of the gas, indicating that the monsoon was sufficiently powerful to breach the tropopause and create a superhighway skyward for all kinds of atmospheric nasties.

'The monsoon is one of the most powerful atmospheric circulation systems on the planet, and it happens to form right over a heavily polluted

*   Science 30 April 2010: Vol. 328, no. 5978, pp. 611–13
     DOI: 10.1126/science.1182274

region,' says Randel. 'As a result, it provides a pathway for transporting pollutants up into the stratosphere.' So in the monsoon, Asia effectively has its own atmospheric extractor hood directly overhead.

For the moment, researchers don't know what impacts this fast-track route to the stratosphere could be having, but they're worried because once they reach the stratosphere, the kinds of chemicals being carried aloft like this can continue to circulate around the globe for years. This suggests that the impact of Asian pollutants could increase significantly in the coming decades owing to intensifying industrialisation, especially in China. Another uncertainty is how climate change could impact on the power of the monsoon, which could accelerate all of these effects. Never underestimate the power of the wind, it seems – especially on Boxing Day . . .

FACT BOX

What else can the wind do?

Apart from providing Asia with an atmospheric equivalent of a giant vacuum cleaner to eject pollution skywards, scientists have also uncovered evidence recently that large storms might make the earth move too, by setting off certain types of earthquakes.

The phenomenon was discovered by Taiwanese researcher ChiChing Liu and two US scientists, Alan Linde and Selwyn Sacks.[*] Between 2002 and 2007, they buried underground strain-sensing devices in eastern Taiwan and used them to follow how the earth's surface was deforming in this region over time. They were quite shaken to discover that these devices could actually pick up the arrival of the big seasonal tropical storms called typhoons, which occur predominantly in the second half of each year.

These storms are accompanied by very low

[*] *Nature* 11 June 2009: Vol. 459, pp. 833–36
DOI: 10.1038/nature08042

pressure, which usually causes the ground to swell, which is what the team could see on their subterranean strainmeters. But occasionally they would pick up the reverse effect – the ground appeared to have shrunk rather than stretched when a storm came.

In all, they detected 11 events like this, each of them associated with typhoons. The likelihood of this occurring by chance is less than one in a million and the only explanation, say the scientists, is that the typhoons are triggering 'slow earthquakes', which are ground movements that occur over longer time scales – hours to days – than their normal vigorous counterparts, which tend to subside within seconds.

The storms unleash the slow quakes by increasing the stress across faults. This occurs because the arrival of a storm system causes the pressure to drop over the land surface, but to remain unchanged over the sea. This is because low air pressure occurring over the sea causes more water to move into that area to compensate, so the sea floor feels the same amount of force. As a result, during a storm the stress across nearby faults can increase and if they are primed to move, a quake follows.

Paradoxically, this mechanism might actually help to protect Taiwan, where the Philippine Sea plate and the Eurasian plate are colliding. The plates are trying to move past one another at more than eight centimetres per year, so by periodically 'unloading' the fault, these slow quakes can prevent a build-up of energy that would otherwise be unleashed suddenly with potentially catastrophic effects, as has occurred with devastating effect in Iran and Haiti recently.

Will there be an earthquake 'toad-day'?

The prevailing view about whether it's feasible to forecast the date of an earthquake is that you can't. The best we can do, scientists say, is to try to spot which faults – the locations where two or more tectonic plates touch – are storing up energy and trouble for the future, and then warn people locally, although when the fault will finally give way it's impossible to say. But now, new research suggests that this is something of a seismic myth, because toads, it turns out, appear to be the quake-equivalent of a caged canary – they can tell well in advance when they need to hop it!

All around the globe, geologists are doing the hi-tech equivalent of holding a glass to the wall to pick up on the inner murmurings of the earth in order to identify signature sounds that might herald the arrival of a forthcoming earthquake. At the same time, other scientists are scrutinising the planet's surface in excruciating detail from space, looking for subtle signs and deformations in the lie of the land which might indicate that an upheaval is imminent. This is important because unearthing evidence that a quake may be about to happen is a major scientific priority. According to the United Nations, earthquakes cause an average of 78,000 deaths worldwide per year and have accounted for 60% of all natural disaster deaths since the year 2000.

As geologists are quick to stress, it's not actually earthquakes that usually kill people: collapsing buildings do. Therefore, knowing when one is due to happen in advance could make a life and death difference for those affected, because it would enable people to be safely evacuated beforehand. This would be especially helpful in developing countries, where the buildings are of generally poorer quality and usually aren't designed to be earthquake-resistant. That said, despite years of effort, and the fact that scientists now know where most of the likely trouble spots are, researchers aren't much closer to telling exactly where and when disaster will strike. It's quite surprising then to find that the humble toad seems to be able to tell what geologists can't – and presage a quake more than five days ahead.

Two scientists from the UK's Open University,

Rachel Grant and Tim Halliday,[*] uncovered evidence of this extraordinary amphibian example of clairvoyance while monitoring mating toads around the San Ruffino lake in central Italy during the spring of 2009. Usually, toads converge on the lake in large numbers and time their matings to coincide with the full moon. They then remain in the vicinity for a month or two, until spawning is complete, before dispersing again.

But on 31 March 2009, when breeding had barely begun, the 90 or so animals that had gathered to mate all abruptly vanished. Six days later, the area was hit by a magnitude 6.3 earthquake. Some toads did return at the full moon shortly after to mate, but the majority remained absent for a further nine days until the last significant aftershocks had finally died away. Then their numbers climbed again. Eliminating other major factors, like the weather – which didn't change – the only explanation is that the toads somehow detected the impending earthquake five days ahead of its arrival.

Incredible as this sounds, the results resonate with other previous studies that indicate certain

* _Journal of Zoology_ August 2010: Vol. 281, no. 4, pp. 263–71 DOI: 10.1111/j.1469-7998.2010.00700.x

species might be able to predict when a seismic shake-up is on the way. In China, scientists reported in the 1970s that fish, rodents, wolves and snakes had been behaving oddly up to two months before a major 7.8 Richter scale quake in Tanshang. There are also reports going back to the 1920s from Europe, Asia and the Americas describing strange behaviour in fish, rodents and some snakes up to a week or so prior to a quake.

So the effect does appear to be real and probably reflects a survival strategy on the part of the species concerned. The big question is, how are these toads telling when it's time to go? For now, scientists don't know, but there are several possible ways it could happen. One is that the toads could be detecting transient peaks in the levels of radon gas. This is a natural by-product of the radioactive decay of uranium in subsurface rocks and it can build up in pockets underground. These can be released by earth movements, which open up cracks, helping the gas to escape to the surface like a geological belch. Since toads are very sensitive to the chemistry of the local aquatic environment, perhaps a spike in radon levels could be sufficient to make them jumpy?

Another possibility is that local changes in

the planet's magnetic field, provoked by a fault gearing up to go, could upset some animals. Toads have an inbuilt biological magnetic compass to help them navigate, so if an impending upheaval alters the magnetic field, this could perturb their behaviour. Some scientists have also suggested that leading up to an earthquake, changes might also take place in the planet's ionosphere, the blanket of charged particles that envelops the earth out in space. This can be monitored using very-low-frequency (long-wave) radio signals that can bounce off the charged layer. Intriguingly, records from the area in the spring of 2009 show that there were indeed disturbances in the ionosphere from five days before the Italian quake occurred, although quite how animals might pick up on this nobody knows.

But regardless of how they're predicting the future, these amphibians have clearly worked out a way to avoid croaking it – so if you're off to an earthquake-prone area, take a toad. If he legs it, so should you!

Is love really an addiction?

'Love is the drug and I need to score,' sang Bryan Ferry in the 1970s, earning him a smash hit and a small fortune. But apart from being a catchy song lyric and a cliché, that line is also looking less like a myth and more like a scientifically accurate fact of life. That's because, in recent years, researchers have begun to bring the power of modern genetics and neuroscience to bear on the workings of the human psyche, including the 'big' question of love and what it is.

Somewhat unromantically, the results of these endeavours are showing us that the simple answer is that love amounts to little more than a chemical addiction. In fact, the same brain circuits become active when volunteers in a scanner are shown pictures of their loved ones as when a nicotine-starved smoker lights up their first cigarette of the day. The molecular clockwork of that lovin' feeling, it turns out, is a small family of nerve transmitter chemicals called oxytocin, vasopressin and dopamine. Oxytocin is released in the brain during orgasms, during childbirth and by breastfeeding, which has led scientists to suspect that it may be linked to mother–baby bonding and that perhaps this, 'love' and partner attachments are all a manifestation of the same process.

Experimentally, the evidence so far is quite compelling. Amongst sheep, a mother can be persuaded to foster a lamb that isn't her own by delivering a brief puff of oxytocin into her nose before introducing the baby. Unmated female rats also become highly maternal around rat pups that they would previously have killed when given a dose of the chemical beforehand. Humans are affected, too. Volunteers given doses of oxytocin develop enhanced sensations of trust for those nearby, become more sensitive to the emotions of others and spend longer looking at people's faces (which is a first for many blokes). This suggests that couples who experience orgasms together are effectively programming each other's brains to love and trust one another.

But trust also usually demands monogamy, the mediator of which is vasopressin. Studies on voles have shown that a polygamous vole species known as the meadow vole can be transformed into behaving like its monogamous prairie vole cousin either by adding extra vasopressin to its brain, or by increasing the brain's sensitivity to the substance. The same seems to apply to humans, because a study carried out in 2008 in Sweden found that individuals with one variant of a gene used in the brain to detect vasopressin levels were twice as likely to report a recent marital crisis, and only half as likely to be married in the first place, compared with individuals not carrying that form of the gene.* Administering vasopressin to volunteers also produces changes in behaviour. Men adopt a more aggressive posture, which includes looking more menacing and becoming much more protective of their partners. When shown photographs of other people's faces, they tend to rate them as looking less friendly than they did before vasopressin was given.

So what about the addictive part of love: the sensation that you cannot survive without the other person, and the rush of joy when you see them after being away? This is down to dopamine, the brain's pleasure chemical. When nerve cells squirt small amounts of this into a brain region called the nucleus accumbens, it

*   *Science* 7 November 2008: Vol. 322, no. 5903, pp. 900–04
DOI: 10.1126/science.1158668

produces sensations of euphoria and satisfaction. We use this circuitry to reward ourselves when we do something right, whether that's learning a new fact, passing a driving test or making someone happy. It's the way that the brain reinforces learning and good behaviour.

Stripping Down Science

It's also the target of drugs like cocaine and heroin, which effectively short-circuit this same brain mechanism to achieve their pleasurable effects.

This is where Bryan Ferry's famous lyric comes in, because dopamine lies downstream of the effects of the other two chemical love-drugs, vasopressin and oxytocin. When these chemical signals are active, they trigger the release of an addictive surge of dopamine in order to consolidate their effects. So you are, quite literally, getting hooked on your partner.

Being able to distil love down to a series of chemical reactions like this is informative and helpful on the one hand, because it will very likely enable scientists and doctors to help patients with conditions like autism, which make it hard for them to form relationships with other people. But it also opens the door to a much more nefarious future, in which we will have the pharmacological ability to manipulate love with a drug. For now, though, the chat-up line 'Could you just sniff this?' should probably serve as a warning . . .

Planets don't alter their orbits, do they? Or is there a gap in our celestial knowledge?

When scientists first theorised how the solar system formed, the view was that the planets didn't move much from the orbits they occupy today. More recently, people have come round to the idea that this is probably a Jupiter-sized myth and that, in fact, during the early parts of their lives many of the heavenly bodies in our cosmic neighbourhood (Hollywood celebrities aside) were actually playing the planetary equivalent of billiards.

The evidence for these celestial flights of fancy turns out to be lurking in the asteroid belt that lies between Mars and Jupiter. This region of space contains thousands of protoplanetary leftovers from the primordial soup that first spawned the solar system. Essentially, this is material that failed to form a planet and instead exists as a ring of cosmic rubble ranging in size from sub-planet-sized lumps and hunks of rock hundreds of miles across to house-sized pieces and particles smaller than specks of dust.

But contrary to what you might expect, the distribution of the objects in the asteroid belt is not random, something that was picked up on in the 1800s by the American astronomer Daniel Kirkwood. What caught his astronomical eye was that when he looked at the positions of some of the largest asteroids, he noticed that rather like the booze aisle of a supermarket after a ramraid, the belt appeared to have been emptied of objects in certain places.

Kirkwood put this down to the immense sizes of Saturn and Jupiter, suggesting that, fed by a massive gravitational appetite, these two giants had plucked objects out of the belt at locations where, like two bells chiming together and resonating, the orbital periods of the two planets and the missing material were whole-number multiples of one another. These gravitational resonances, he argued, would have ejected material from certain parts of the belt, leaving empty space. In his honour, these asteroid holes were subsequently christened 'Kirkwood gaps', and over the 150 years that have followed, scientists have gone on to log the locations of all 600 or so of the largest remaining objects flanking the spaces.

Armed with this data, David Minton and Renu Malhotra, two researchers at the University of Arizona,[*] wondered whether the locations of all of these objects, now they are known, would still square with Kirkwood's original explanation. To test the theory, they built a computer model of the early solar system, placing the planets where they presently are and stuffing the asteroid belt full of material to start the planetary balls rolling. They then allowed their model solar system to evolve inside the computer for the equivalent of several billion years to see whether the end result would resemble the real-life configuration out there in the asteroid field today.

What the computer produced was a pattern of gaps that lined up to a certain extent with what we see in reality, but not quite. For a start, there were far more asteroids remaining in the simulation than there should have been. Looking more closely, Minton and Malhotra found that

[*]   *Nature* 26 February 2009: Vol. 457, pp. 1109–11 DOI: 10.1038/ nature07778

on the inner edge of the simulated asteroid belt, facing the sun, there were holes that matched up with the Kirkwood gaps. But on the outer edge, furthest from the sun, the gaps were still occupied by asteroids. To resolve this cosmic conundrum, the scientists tested what would happen if, rather than keeping them in one orbit, they allowed Jupiter and the other giant planets to move into closer or more distant orbits as the solar system was forming.

This proved to be the eureka moment, because when they re-ran the simulation with Jupiter starting off slightly further from the sun and then migrating inwards over time, while Saturn, Uranus and Neptune all moved out slightly, the gaps missing from the model asteroid belt suddenly appeared. In other words, these planetary movements produced just the right patterns of gravitational resonances over time to dislodge the extra asteroids. So it's slightly ironic that by looking for something that isn't there, scientists can retrace the steps taken by the biggest planets in the solar system over four billion years ago. As David Minton puts it, 'The patterns of depletion are like the footprints of wandering giant planets preserved in the asteroid belt.'

So what would have happened to the asteroids that were pulled out of the belt? Well, it's likely that they were the cause of a violent period in the solar system's early history called the late heavy bombardment, which occurred about 3.9 billion years ago when the earth was just over half a billion years old. During this time, which lasted for a few hundred million years, large numbers of objects rained down on the planets in the inner solar system, slamming into them and producing enormous craters. Earth's moon still bears the scars on its surface today.

But why would Jupiter and its gas-giant accomplices have wandered about in the first place? David Minton speculates that Saturn, Uranus and Neptune all moved outwards because they interacted with a massive disc of material on the outer edges of the solar system called the Kuiper Belt. By dislodging large bodies here, the planets would have been pulled steadily outwards until the Kuiper Belt was depleted of material, bringing the outwards migration to a stop.

And Jupiter? Interestingly, the answer to this puzzle probably lies outside our own solar system. In recent years, astronomers have discovered Jupiter-sized planets orbiting distant stars but, in many cases, these so-called 'hot Jupiters' are so close to their parent stars that they couldn't possibly have formed in that position, meaning they must have migrated there. Researchers believe this happens because the planets pick up progressively more material lying between them and the star, causing them to be drawn inwards towards the centre of the solar system. Luckily for us, our own Jupiter didn't move in too far, or we would have been on a collision course with a juggernaut.

FACT BOX

Clever way to spot alien planets

Because cosmology and space science tend to happen on time scales that are measured in billions of years, it can sometimes be easier to answer questions about the past and future of our own solar system by looking at a different one that happens to be the age you're interested in. But in 2003, the International Astronomical Union, meeting in Australia, estimated that there are at least 70 sextillion (seven followed by 22 zeros!) stars out there in the universe, and at least 200 billion just in our own Milky Way Galaxy. So how do we know which ones to look at, especially if we want one with planets we can study?

Previously, it was a painstaking task of watching stars carefully to see if they wobbled a tiny bit or their light altered slightly, which could indicate a planet was in orbit. But in 2009 scientists made a breakthrough that should speed up the whole process. Garik Israelian, from the University of San Fernando de La Laguna in Tenerife,* surveyed more than 500 stars (of the celestial variety), 70 of which have already been confirmed as having planets in orbit around them. By analysing the spectrum of light from each star, it was possible to measure, using a technique discovered in the 1800s by none other than Robert Bunsen, what chemical elements were present.

When the data from these stars were

* *Nature* 12 November 2009: Vol. 462, pp. 189–91
DOI: 10.1038/nature08483

compared and factors such as age were taken into account, a surprising trend emerged. The majority of the stars showed evidence for the presence of the element lithium on their surfaces, but the 70 stars *known* to have planets orbiting them, our own sun included, had hardly any.

Israelian thinks that the presence of planets somehow stirs up the substance of the star, pulling any lithium on the star's surface into the hot interior where it is consumed. How exactly this happens no one yet knows, but the key point is that, by looking for a lack of lithium in the spectral signature coming from a star, space scientists now have a shortcut way to find new planets much more quickly – so watch out, ET, here we come!

People don't really walk in circles, do they?

According to Mark Twain in *Roughing It*, JRR Tolkien in *The Lord of the Rings*, and various picture house hits including *The Blair Witch Project* and *The Flight of the Phoenix*, people are prone to wander around in circles when they get lost – and when they do it on camera, they also inevitably wind up with an Oscar and a box office blockbuster!

Previously, science has had very little to say about this claim, which has been dismissed by many as mere Hollywood hype. But now scientists have put this potential myth to the test and, it turns out, this is one circular argument that does have some facts to back it up.

Jan Souman, a researcher at the Max Planck Institute in Tubingen, Germany,[*] unleashed six walkers in unfamiliar forest terrain and told them to walk in a straight line for several hours.

[*]   *Current Biology* 20 August 2009: Vol. 19. no. 18, pp. 1538–42
     DOI: 10.1016/j.cub.2009.07.053

He also repeated the process with three subjects who were asked to walk in the Sahara Desert. The progress of all of the walkers was charted using a GPS system to follow where they went. In both geographies, the walkers went off course whenever they couldn't see the sun (or the moon, in the case of one nocturnal Saharan rambler). Under these circumstances, the routes they took quickly became circular, with the walkers frequently crossing and re-crossing their own paths.

But why? Haven't we got a sense of direction? One possibility is that this happens because, just as we have a tendency to favour one hand or one eye over the other, in some people there may also be an innate tendency to turn in one direction. This could also occur for biomechanical reasons, such as having one leg longer or stronger than the other.

To find out whether this was the case, Souman also carried out a series of additional walking exercises on an airstrip (fortunately, one that wasn't in use). With the subjects blindfolded, he told them to keep walking in a straight line in a certain direction indicated to them beforehand. To check for the effects of any biomechanical issues, he also measured muscle power, X-rayed the legs of one individual to accurately gauge their lengths (they differed by less than one millimetre) and repeated the experiment after altering the heights of the soles of the subjects' shoes to make their legs different lengths.

The results showed that there was no correlation between the mean direction in which the subjects turned when they walked blindfolded and any mechanical asymmetry – the subjects just ended up following random trajectories punctuated by small circles which were often small enough to fit inside a basketball court. Interestingly, this meant that the greatest as-the-crow-flies distance any of the subjects got from their starting points during over 50 minutes of blindfolded walking was just 100 metres. Helpfully, this suggests that if a person was lost and blundering about in dense forest without access to any visual landmarks, concentrating a search within the area in which a missing person was last seen would be the best strategy for rescuers.

Why does this happen? Souman thinks that people veer off course in the absence of a visual guidance cue, like the sun, because of 'neurological noise' – small errors in the processing of the motor, sensory and balance systems in the brain, which alter the subjective sense of what is 'straight'. Quite literally, and with every step, a random error is added to the subjective 'straight ahead', causing it to drift off true. When these deviations become large, people often end up walking in circles, regardless of whether they have a good 'sense of direction' or not.

The moral of this story is, when lost in the forest – or the Sahara – follow the sun or the moon, or stay put!

FACT BOX

Human navigation

Although Jan Souman's study suggests that we have an innate tendency to become lost, many people nonetheless claim to have a strong sense of direction and can tell which way they ought to be going. But is it real and can it be relied upon?

This was looked into by a Manchester University scientist, Robin Baker, in a study he published in 1980.* He blindfolded 64 student volunteers and drove them along a tortuous route to a series of remote locations up to 52 kilometres from the university. At each location, the volunteers were asked – without removing the blindfold – to indicate the compass direction in which they thought the university lay. In each case, the results were strongly correlated with the real direction, suggesting that the subjects innately appeared to know where home was situated, despite none of them being able to account for how they arrived at their directional decisions.

To find out whether they might be tapping into the earth's magnetic field as a form of cognitive compass, 15 of the subjects had bar magnets strapped to their heads. The rest of the group were given identical non-magnetic pieces of metal, but no subject knew whether they had a real magnet or not. The results from the magnet wearers were all well off-course compared with their control colleagues,

*   *Science* 31 October 1980: Vol. 210, pp. 555–57
     DOI: 10.1126/science.7423208

suggesting that magnetic cues could be playing a part in helping to drive our sense of direction. As yet though, no one has found what underlies this ability, or even which part of the brain, head or neck is responsible.

But if we have an inbuilt direction-finding system, why do people wander in circles, as Souman says? Probably because, as humans, we have evolved to set higher store by another, more dominant sense. Over a third of the human brain is devoted to processing what we see so, not surprisingly, we tend to focus on what our eyes are telling us and prioritise this over other subtle cues and signals coming in from the world around us.

So how *do* we find our way around, or recall the route we took across town, or retrace our steps to find the restaurant where we left the umbrella at lunchtime? Well, scientists have discovered in recent years that the brain uses a neurological grid system resembling a three-dimensional radar screen with us plotted in the middle of it. The grid is located in a part of the brain's temporal lobe called the entorhinal cortex and it consists of an array of interconnected nerve cells linked up to form a series of equilateral triangles. When a person moves, their blip on the radar screen is tracked by altering the firing activity of the nerve cells that form the region of the grid representing the part of the world in which they are standing. In this way, the person knows where they are in their environment at any given time.

The first insights into the workings of this system were provided by making recordings from individual nerve cells in the brains of rats and mice as they foraged for food. Scientists realised that these animals were finding their way around by orientating their movements relative to various landmarks that were also plotted on this neurological grid. If those landmarks moved – for instance, if a rat's cage was turned 180 degrees – then the grid would reconfigure to reflect the new locations of the landmarks relative to each other.

Of course, it's not practical or ethical to implant electrodes into the heads of humans for the purposes of monitoring movements.

However, it is possible to use brain scanners, and a researcher at University College London, Christian Doeller,* has recently developed a computer-based system to confirm that the human navigational system seems to work the same way as a rodent's. Forty-two human volunteers were brain-scanned as they explored a 3D environment shown to them through a virtual reality headset. Incredibly, the brain scanner was able to pick up in the subjects' entorhinal cortices the same neurological signature that would be expected based on the workings of the equivalent region of a rat's brain.

Thankfully, the similarity doesn't appear to extend to experiencing a compulsion to dive down the nearest sewer, although some people do rummage in dustbins – perhaps we now know why!

* *Nature* 4 February 2010: Vol. 463, pp. 657–61
DOI: 10.1038/nature08704

Baby brain drain: does pregnancy kill your IQ?

It's commonly claimed that for women, pregnancy and childbirth shrink your brain and erode your memory. In fact, anyone who reads just a fraction of what's penned in the birthing literature about what some are calling 'placental brain drain' would be excused for thinking they should be checking themselves into a dementia clinic rather than a maternity hospital! But is there really any reliable evidence to back up these claims of baby-induced intellectual meltdown? Surprisingly, or perhaps not, the answer is actually 'no', and recent research is now suggesting that the whole thing may just be an over-gestated myth well past its due-by date.

So how were these ill-founded claims of compromised cognition conceived in the first place, and how have they managed to implant themselves in our psyche? It probably all stems from the fact that the majority of the studies that have looked at women's brains during pregnancy have taken small groups of pregnant women and compared them with other small groups of non-pregnant 'control' women to see how the cognitive abilities of each match up.

The obvious problem with this sort of study design is that the one thing you're seeking to find out – whether the brainpower of the pregnant person has changed – isn't being tested at all, because the pregnant subjects were invariably not assessed before they became pregnant to find out what was 'normal' for them. At the same time, many of the studies relied on participants self-reporting symptoms of memory loss or poor concentration. And because pregnancy is stressful at times, and pregnant women have heard the claims that their IQ ought to be dropping faster than a faulty facelift, so-called 'recall bias' kicks in and a self-diagnosis of borderline idiocy is made.

To resolve this problem, three Australian researchers – Helen Christensen, Liana Leach and Andrew Mackinnon[*] – have now carried out a longitudinal study in which participants are followed up over a long period of time to look for

*   *British Journal of Psychiatry* 2010: Vol. 196, pp. 126–32
     DOI: 10.1192/bjp.bp.109.068635

'before and after' changes in relation to various lifestyle factors, including having children. The Australian team contacted 1241 women who had been enrolled in a study called the Personality and Total Health (PATH) Through Life Project, a community survey concerned with health and wellbeing. These subjects had joined the study in 1999 when they were between 20 and 24 years of age. At this time, baseline measurements of cognitive speed, working memory and immediate and delayed recall abilities were collected. The women were then reassessed using the same parameters in 2004 and 2007, five and eight years later.

Over this timeframe, more than 250 of the enrolled women had become pregnant and had children, which meant that Christensen and her colleagues could directly compare an individual woman's cognitive performance before, in some cases during and also after pregnancy. Reassuringly, the study didn't find any significant pre- and post-pregnancy cognitive differences between the participants in any of the tests. One significant difference was detected – a small drop in cognitive speed amongst women who were pregnant at the time – but this only emerged when the pregnant cases were split up into two subgroups – early and late pregnancy – which means that it could be statistically unreliable.

Instead, it's much more likely that women succumb to dodgy dogma and think themselves into a cognitive corner while they're pregnant, rather than suffering any real reduction in mental processing power. This was also the conclusion of another researcher from the University of Sunderland, Ros Crawley.* She asked a mixture of people, including men and women, both with and without children, how they thought women

*   *Applied Cognitive Psychology* December 2008: Vol. 22, no. 8, pp. 1142–62 DOI: 10.1002/acp.1427

were affected by pregnancy. Amongst all the groups who replied, the responses indicated the clear belief that cognitive abilities decline during pregnancy. She also asked pregnant women to rate whether they felt their cognitive prowess, including their driving abilities, had been dented during gestation. This elicited a significant 'yes' response compared with non-pregnant women.

Predictably, however, when she tested the same subjects using memory tests and a driving simulator task during which they had to react quickly when a vehicle in front stopped, or pull out safely from a junction into a stream of traffic, the pregnant women performed as well as the non-pregnant participants. (Although, admittedly, driving along with the indicators on or the handbrake still applied were not assessed.)

The moral of the story is that it's probably not the pregnancy brain drain that we should be worried about so much as the wallet drain that inevitably ensues once the baby is born . . .

Forgetful fish, or 'carp-acious' memory?

It's often said that fish have a microsecond memory, with every lap of the tank representing a fresh foray into uncharted waters. Nor does Disney do fish any favours with its box office smash *Finding Nemo* telling the story of Dory, a Regal Tang with a memory problem. However, a trawl through the published literature reveals that fish probably have much better memories than we first thought, and are also adept at social networking, educating each other, slipping through nets, categorising music and even learning online (on fishing line, that is, because after they've been hooked once, it turns out that they're much less likely to make the same mistake again, assuming they survive the experience).

One very elegant experiment to showcase these skills in action was carried out by Culum Brown, a marine researcher now based at Macquarie

University in Sydney.[*] He engineered a net that contained an escape route. This was swept along the length of a tank housing groups of five spotted rainbowfish, which had been collected previously from Amamoor Creek in Queensland. Over a series of trials, he timed how long it took the fish, once they entered the net, to find the hole and to escape. After just five goes at the task, which took less than 15 minutes, the fish had almost halved the time it was taking them to escape, which is hard to reconcile with an allegedly cerebrally challenged animal.

Even more convincing was that when the same fish were re-tested 11 months later, having not seen the apparatus in the interim, they all escaped as quickly as they had when fully trained previously, indicating that they must have remembered the way out from before. For a species that only lives for a few years and apparently has a memory deficit, that's pretty impressive recall!

Fish also appear to have a musical side, as revealed by Ava Chase, from the Rowland Institute for Science in Massachusetts, US,[**] who

[*] Brown et. al. (Eds) (2006), *Fish Cognition and Behavior.* Wiley-Blackwell. DOI: 10.1002/9780470996058

[**] *Animal Learning & Behavior* 2001: Vol. 29, pp. 336–53

successfully trained three carp to discriminate between blues and classical music played through a speaker submerged in their tank. Ironically, the blues tracks she chose were by John Lee Hooker, but this still proved highly catchy with the three koi study subjects – Beauty, Oro and Pepi – who could all accurately tell, by pressing a button with their mouths in return for a food reward, classical genres (including both baroque and non-baroque composers like Mozart, Beethoven and Schubert) from blues numbers. Incredibly, the fish could also correctly categorise music they'd not heard previously, although they did initially struggle with a series of Vivaldi guitar concertos. This was remedied by switching on a sound filter to screen out low frequencies around 200 Hz, which seemed to be distracting the fish, possibly by stimulating their 'lateral lines', sense organs which pick up low-frequency vibrations from water.

This sensitivity of fish to sound has led some scientists to investigate the possibility of using sonic techniques to make them easier to catch. Plymouth University psychologist Phil Gee[*] has recently been working with the UK's National

*   www.dailyrecord.co.uk/news/uk-world-news/2008/02/02/pavlov-experiment-shows-fish-are-not-forgetful-86908-20306570/

Marine Aquarium to train Bentley, a metre-long, six-year-old giant humphead wrasse, to swim into a special holding area connected to his main swimming tank at meal times. The dinner bell consists of simply banging on the door of the tank, and the fish responds because it has learned, rather like the piscine equivalent of Pavlov's dogs, that it gets a reward whenever it cooperates. As further evidence that fish don't forget, when the training schedule was interrupted by maintenance work being carried out on his tank, Bentley still performed faultlessly after a gap of four months.

Phil Gee has also reported previously, using goldfish from a garden centre, that the animals can be trained to press a lever to release food rewards. Even more cunningly, when the lever was rigged to feed the fish only at a certain time of day, the fish all learned to press the lever solely when it would reward them, indicating that they can also keep track of time.

This tallies with what fish owners report themselves, which is that fish in home aquaria and ponds seem to know when it's feeding time and swim to the side of the tank in anticipation of a meal. As a result, researchers are now testing whether it's possible to use these observations to set up free-range pisciculture (fish farming) systems which can rely on sounds to 'call in' the fish at feeding (and harvesting) time. This could save Third World fishermen a fortune in expensive equipment if it can be made to work. So far, tests have taken place in Norway and a pilot study has been set up in Ghana. There's been no word yet, however, regarding whether this scientific splash will actually work or just turn out to be a belly-flop . . .

So, far from being forgetful, fish appear to have a 'carp-acious' appetite for learning and memory and even a taste for music. Whether it goes as far as 'back to my plaice' for some 'drum' and 'bass' remains to be seen!

'TACKLE' BOX

Ecological significance of fish memory

Apart from being able to spot when it's dinner time or remember the best places to hide in their home ranges, researchers have also shown that fish do most of their learning by watching each other. This was demonstrated recently by Durham University scientist Jeremy Kendal[*] using 270 sticklebacks (known non-scientifically as 'tiddlers') he had caught in a local river in Leicester.

In the laboratory, small groups of the fish were placed in a tank equipped with feeders at either end, one of which dished out generous portions of bloodworms, which the fish love to eat, while the other was much stingier in its offerings. The fish quite quickly learned the drill, showing a strong preference for the more generous feeder. At this point, they were then confined within a 'viewing gallery' section

[*] *Behavioral Ecology* 2009: Vol. 20, no. 2, pp. 238–44
    DOI: 10.1093/beheco/arp016

of the tank while a new group of fish were introduced.

As an added twist, this time the scientists reversed the feeders so that the previously generous one was now stingy and vice versa. While this was going on, the confined 'educated' sticklebacks were left to watch how the newly introduced fish fared. Afterwards, the viewing fish were released again so the team could observe which of the two feeders they now favoured.

Incredibly, just by watching how the other fish had got on, over 75% of the observing fish had learned that the feeder situation had reversed, which they demonstrated by making a beeline straight to the opposite end of the tank. More impressive still was that in a subsequent experiment when the team adjusted the relative generosity of the feeders, the observing fish changed their behaviour only if they saw the other fish doing better in the new situation than they had in the old one.

This kind of learning approach, known as a 'hill climbing strategy', is probably key to helping these animals to escape from predators. 'These fish are too vulnerable to forage alone,' explains Kendal, 'so they have to move around in groups. They are therefore social, and by watching the outcomes of others, and responding appropriately, this is a sound strategy to avoid predation and maximise returns.'

This social learning strategy also suggests that, in the same way that human culture passes down knowledge and learning – such as how to read, write and count – from one generation to the next, fish and other animals may have a similar set-up. Knowledge of the best feeding grounds, the best breeding grounds and the best migration routes is perhaps actually being passed non-genetically amongst members of a population.

This could partly explain, Culum Brown points out, why Atlantic stocks of cod have collapsed so severely – because the big fish with all the know-how have ended up on people's dinner plates, taking their life experiences with them.

Stone the crows! Aesop was right

One of Aesop's famous fables tells of a thirsty crow that, unable to push over a pitcher of water or reach down to the liquid inside, drops in stones until the level rises to within a beak's-length of the top, enabling the bird to take a refreshing drink. The story seeks to emphasise the importance of brains over brute force and was thought to be just that – a story. But now scientists have shown that at least one member of the crow family really can do this, proving that it's a myth that this Aesop's fable is a fable.

In what turns out to be an amusing example of nominative determinism, Cambridge University researcher Christopher Bird and his colleague Nathan Emery[*] challenged four peckish rooks (relatives of crows) called Cook, Fry, Connelly and Monroe, to serve themselves a tasty treat in the form of a worm floating out of reach in an upright tube of water. The birds needed to work out how to raise the water level in the tubes by

*   *Current Biology* 6 August 2009: Vol. 19, no. 16, pp. 1410–14
DOI: 10.1016/j.cub.2009.07.033

shovelling in stones that were lying nearby to bring the worm within pecking distance. Although none of them had seen this precise experimental set-up before, and this particular species is not known for using tools in the wild, they all quickly solved the problem.

In an added twist, when the researchers varied the water level in the tubes they found that the birds would initially size up how many stones they thought they would need and first drop in that number before making any attempt to retrieve the worm. Also, when the four rooks were offered a choice between using large and small stones, they invariably selected the large ones, having realised that these raised the water level more quickly. The researchers point out that the birds never added further stones to the tubes once they had successfully plucked out the worm, indicating that they genuinely saw the stones as a means to a meal, rather than performing the act for any other reason.

To prove that the birds were really learning from and reacting to the results of their efforts, the researchers included a tube containing sawdust in place of water. The rooks quickly realised that in this setting, the stones made no difference to the height of the worm, so they stopped dropping them in. This shows that just like their close relatives the crows and jays, which have been labelled by some as 'feathered Einsteins', rooks also possess a remarkable aptitude for problem solving. They owe their intellect, as do other corvid (crow) family members, to their large brains which, relative to their body sizes, put them at par with a chimpanzee.

One other unusual feature of this bird family is that they possess the rare (amongst animals) ability to recognise themselves in a mirror. Most animals – human babies included – when presented with their own reflections are fooled into thinking they are seeing another individual and respond with displays of interest or aggression. Despite multiple exposures, they never seem to learn that they're staring at themselves. But so-called 'higher' animals, including chimps, dolphins and elephants, don't fall for this trick and neither, it now turns out, does another crow relative, the European magpie. This discovery was made recently by Helmut Prior, from Goethe University in Germany.[*]

* *Nature* 22 February 2007: Vol. 445, pp. 919–21 DOI: 10.1038/nature05575

He made small red or yellow marks on the neck feathers of Lilly, Harvey, Gerti, Schatzi and the somewhat oddly named Goldie, five magpies in his laboratory. When the black-and-white birds were placed in front of a large mirror, they scratched at the marks on their feathers. As soon as they successfully removed them, the scratching stopped, indicating that they realised that the reflection was their own image and suggesting that they may have what scientists call a 'theory of mind'.

Other experiments have shown that animals of this family also have a strong sense of time and can remember the past and use that knowledge to plan for the future. Working with scrub jays, Cambridge scientist Nicky Clayton* built the crow equivalent of a hotel, with a dining room where 'residents' were always fed and a sleeping room where they were locked at night. After a few stays in the hotel, the birds realised that once they were shut into the sleeping area, there was no further access to food. Their solution? On subsequent occasions, as soon as they were fed in the dining room, they would sneak into

*   *Science* 9 August 2002: Vol. 297, no. 5583, p. 981
DOI: 10.1126/science.1073433

the sleeping area and hide a tasty treat or two, on the off chance they might be in the mood for a midnight feast later!

Some crow family members are also known to be impressive toolmakers, but researchers at Oxford University were still gobsmacked when a New Caledonian crow they were studying called Betty picked up a piece of straight wire and bent it into a hook in order to retrieve an object lodged at the end of a pipe. Alex Kacelnik and his colleagues[*] had given Betty, and a second male crow called Abel, a choice between either a straight or hook-shaped piece of wire to see which they would use to retrieve a snack. But when Abel got in a flap and stole the more useful hook-shaped wire, Betty made her own. According to Professor Kacelnik, 'Although many animals use tools, purposeful modification of objects to solve new problems, without training or prior experience, is virtually unknown'.

But the icing on the cake must be the canny crows filmed in Japan by David Attenborough for his *Life of Birds* television series. In this instance, crows in Tokyo had learned to make use of

[*]   *PLoS Biology* August 2008: Vol. 6, no. 8, e202
      DOI: 10.1371/journal.pbio.0060202

traffic lights and pedestrian crossings across the city to unlock a new source of food – previously impenetrable nuts. Having collected a kernel, a bird would drop it onto the road into the path of the oncoming traffic, close to a crossing. It would then wait patiently for a car to crush it open. Once this had happened, and at the next convenient gap-in-the-traffic juncture – usually when the lights were red, the crow would hop down and retrieve its titbit. Initially just a few of the animals were at it, but quickly they learned the trick from each other until it was common crow knowledge!

FACT BOX

Don't mock my memory

It's not just crows that are endowed with an impressive intellect: other birds have been shown to possess remarkable memories too, including mockingbirds.

In a recent study, University of Florida

Gainesville researcher Douglas Levey[*] asked a human volunteer to repeatedly disturb a pair of nesting mockingbirds over several days while he monitored the birds' responses. (Sensible chap. If I were the volunteer, I'd have asked to swap!) Predictably, with each successive incursion close to their nest, the birds made increasingly exaggerated responses, including making more frequent attacks on the approaching human on day four compared with day one.

One might predict, therefore, that a second – unknown – human approaching the nest on day four would elicit the same vigorous response from the birds, reflecting their generalised state of alarm. Instead, when a second researcher independently approached the nest, the birds reacted relatively mildly as they had done towards the first person the first time he had approached.

This shows that the birds were able to learn very quickly to recognise and distinguish

[*] *PNAS* 2 June 2009: Vol. 106, no. 22, pp. 8959–62
DOI: 10.1073/pnas.0811422106

between different humans who approached their nests. While this sort of behaviour has been seen before amongst social mammals and livestock, and between individuals of the same bird species, birds discriminating between different members of an entirely different species had never previously been reported.

This is important, Levey says, because growing human populations are leading to increased human encroachment into the habitats of many animals. Under these circumstances some species 'urbanise' very well, others less so. The ones that tend to do well under these crowded human-dominated conditions could be those with the sorts of natures displayed by these mockingbirds – the ability to size up, discriminate and react accordingly to different levels of threats, rather than in a one-size-fits-all fashion, which would almost certainly be deleterious to them in the long run – or should that be flight . . .?

Don't talk to strangers

'Don't talk to strangers,' your mother always admonished, along with advice about trusting your instincts and believing in yourself and your own judgement. But this early age brainwashing to ignore the opinions of others in the name of self-belief looks to be bad advice, because a recent study – that had Harvard students speed dating in the name of science – has shown that phoning a friend, or even asking a stranger, will probably provide you with a far more accurate insight into how you will react to future events than if you rely only on your own intuition.

Most of us believe that we are sufficiently well acquainted with our own minds and bodies to make a pretty reasonable assessment of how we'd behave in a given situation. But intuitive and logical as this sounds, surprisingly, it's wrong. Worse still, it appears that we also resolutely refuse to accept the fact.

To prove this, Harvard psychologist Daniel Gilbert[*] set up a study in which he asked 33 female

* *Science* 20 March 2009: Vol. 323, no. 5921, pp. 1617–19 DOI: 10.1126/science.1166632 students to go on five-minute 'speed dates' with one of eight male students. Half of the women, before embarking on the date, were first shown a profile of the man they were going to meet, including his photograph, details about what films, music and books he liked, where he lived and what he was studying. The other half of the women were shown only a report from another woman indicating – on a numerical scale – how much she had enjoyed the interaction with that particular man. Both groups then predicted how much they would enjoy the date, before being taken individually to meet the man in question for five minutes. Afterwards, both women were then asked again to 'rate their dates'.

Incredibly, when the before and after ratings were compared, the women who had been given only the reports of another woman's experience with the same date before making their anticipated enjoyment predictions were found to have been far more accurate than the women who had been given the full profiles of the men before they met them. In fact, the amount by which the women were 'out' in their predictions was nearly 50% lower when they had seen only the report from another woman.

Ironically though, three-quarters of these women nonetheless said that they felt they would have made more accurate predictions about their date if they had seen the full profile.

So does this relate just to a date, or does it apply to other life events too? To find out, David Gilbert set up a further experiment in which he asked a group of volunteers to write a short story which, he told them, would be used to determine which of three personality types they fitted into, A to C, where A was good, B was neutral and C was negative.

In reality, everyone was told they were personality type C, someone who sacrifices their beliefs for an easy life and eschews a challenge.

This was intended to provoke feelings of mild unhappiness amongst the volunteers. Groups of students were then shown either detailed descriptions of the personality types and asked to predict how they would feel if they were rated as each type, or they were shown the response from another volunteer rated previously as personality type C before being asked to predict their own response if they were labelled likewise.

Just like the dating experiment, the students given the reactions of other individuals to look at first were far more accurate – by 63% – in their assessments of how they would feel when they were rated as personality type C, compared with the individuals given the detailed personality descriptions. In this regard, maybe our mothers have 'myth-led' us when they advised against consulting strangers. Instead, they should have referred us to the work of the writer François de La Rochefoucauld, who pointed out in the 17th century, 'Before we set our hearts too much on anything, let us first examine how happy those are who already possess it.'

Despite our hunger for information, knowledge isn't necessarily power, but the opinion of another human being is priceless.

Let it go!
Fingerprints are
not for gripping

For over 100 years, researchers have claimed that fingerprints are there to roughen the skin and help us to get a grip. We're also not alone in having them: koalas, which have a vested interest in holding on tight, are similarly endowed and some South American monkeys possess the equivalent of fingerprints on the tails they wrap around branches to help them hang on as they clamber about in the canopy. Arresting as this argument is, new research has pointed the finger of fate firmly at the door marked 'myth', by showing that fingerprints definitely don't do for dabs what crampons do for climbers!

Having previously got to grips with the physics of fingernails and why they don't split along their lengths (it's because they consist of a laminated three-layered sandwich of keratin proteins, the middle layer of which runs side-to-side across the nail, stopping cracks from propagating wristward), Manchester University researcher

Roland Ennos[*] decided next to grapple with the real role of fingerprints.

Working with his student Peter Warman, the duo designed a 'finger-frictionometer' to measure how much friction the pad of a finger could apply to a sheet of perspex being pulled vertically past it. To simulate different grip strengths or applied pressures, weights were used to accurately alter how hard a lightly clamped finger was pressed against the perspex sheet. Fingerprint impressions were also made in each case in order to measure the area of skin in contact with the surface.

The results convincingly show that fingerprints don't provide a traction boost between finger and surface. This is because, Ennos and Warman found, the skin on the pads of the fingers behaves in a similar way to a rubber ball being dragged across a surface. Unlike 'hard' substances such as rocks, which generate friction when ridge-like irregularities on the surfaces jam together, rubber contains long chains of molecules which form short-lived electrical interactions – called van der Waals attractions – with any surface they touch. So the friction felt when rubber is dragged

*   Journal of Experimental Biology 1 July 2009: Vol. 212, no. 13, pp. 2016–22 DOI: 10.1242/jeb.033977

across a surface occurs because the existing van der Waals attractions have to be broken and new ones made.

This means that, for rubber, the greater the area in contact with a surface, the more powerful the frictional effect. But rubber also deforms when force is applied, which reduces the amount of contact, therefore cutting down the friction, and this is exactly what Ennos and Warman discovered. They also found from their fingerprint impressions that the ridge pattern reduces the contact area of the finger by 33%. So, given that the finger behaves like rubber and that the greater the contact area the greater the grip, if getting a better hold on things is the aim, then it makes no sense whatsoever to reduce the surface area further with a fingerprint.

But if they're not for tightening our grip on things, what are fingerprints for? Some have speculated that the ridges are there for the same reason that tyres have a tread pattern: to channel water away from the contact areas in order to improve grip in the wet. It's also possible that the ridges, which contain vibration-sensing nerve endings, could help to improve the perception of surface textures, although this doesn't explain why parts of the body like the palms and feet, which aren't used for touch discrimination, nonetheless still have ridges.

Ennos' preferred explanation is that the ridges and folds form a clever anti-blister system. 'The pattern will allow our skin to have much greater compliance and that can help to reduce the sheer stresses around the edge of the contact zone,' he says. 'If you ever do DIY tasks, what you tend to find is that the only bits where you get blisters are the bits, not on your fingerprints or where the big patterns are on your palms, but in areas where there aren't any prints.'

In other words, the ridges and folds provide a store of skin that can be stretched out easily like an elastic band to soak up sudden stresses. Unfortunately, however, the system isn't infallible and hammers and thumbs still frequently come to blows!

Dinosaurs were warm-hearted

Most people believe that, in common with the reptiles around today, dinosaurs were cold-blooded brutes that warmed themselves up with the help of the sun. But some hot new research suggests this may be a metabolic myth as massive as some of the dinosaurs were themselves, because scientists have found that, without being warm-blooded, many of these so-called terrible lizards would have moved more slowly than a tectonic plate!

Herman Pontzer, a scientist from Washington University in America, together with John Hutchinson and Vivian Allen from the Royal Veterinary College in London,* made the discovery by using two different scientific approaches to reconstruct the metabolic rates of 14 different dinosaur species ranging from the towering *Tyrannosaurus rex*, right down to the tiny *Archaeopteryx*. Their first method was to take advantage of a relationship spotted

* *PLoS ONE* November 2009: Vol. 4, no. 11, e7783

previously by Pontzer amongst animals around today. When he looked at 28 different land-living creatures, including 11 different mammals, eight birds and five different reptiles, he found that regardless of species, merely by measuring the height of an animal's hip joints above the ground, you can predict with 98% accuracy how much energy they would need to burn to get about.

Suspecting that the same should be true for *T. rex*, Pontzer and his colleagues calculated the hip heights of their dinosaurs and then tweaked the formula slightly to make it more dino-friendly, largely by taking into account differences in posture between modern and ancient species. They then multiplied by one of two speeds, fast or slow, and then by the weights of the animals, based on fossil evidence, to arrive at an estimate for the energy costs of movement for each of the 14 dinosaurs.

To back up these estimates, the researchers then used a second approach, this time based on how much muscle would be needed for each of the dinosaurs to move. This was much more complicated, because it involved modelling how each of the different dinosaurs walked and ran in order to work out how each muscle group would have worked and how much energy it would have consumed. Again, this relied both on fossil evidence, looking at where muscles would have attached to bones, and the muscle mechanics of modern species. Encouragingly, this second set of results agreed very well with the results of the first hip-to-height-based approach.

But the palaeontological problem the researchers then ran into was that the energy requirements they had calculated were nowhere near what a cold-blooded metabolism could muster. Tests on modern-day reptiles running on treadmills, from alligators to iguanas, reveal that these species can only generate energy at the rate of about one-tenth of that achieved by warm-blooded birds and mammals. This means that even for a modest-sized dinosaur, anything more than a sluggish stroll would have been out of the question. In energy terms, the effect would be like trying to light a town with a torch.

So, Pontzer argues, these dinosaurs must have been endotherms (hot-blooded), which is the only way they could have managed to make energy quickly enough to keep them moving. That, or they only went in for sudden sprints of activity followed by a very long recovery period. But this is unlikely, because if something bigger, hungrier and faster came along while they were recovering, they'd be toast – or, at the very least, lunch!

A big question, though, is at what point in their evolution did dinosaurs turn up the thermostat in favour of hot living? Surprisingly, Pontzer thinks they may have started out that way. The 14 examples he examined are positioned in various places – including at the bottom – of the dinosaur evolutionary tree, at the top of which are perched the warm-blooded birds we see around us today.

So, rather than cold-hearted killers, this discovery suggests that being warm-blooded may well have been the ancestral situation for dinosaurs, bringing with it the advantage of being more agile, faster and potentially quicker-witted than their cooler reptilian cousins. This might also go some way to explaining the massive evolutionary success that these creatures enjoyed during the almost 200 million years they ruled the earth.

FACT BOX

New secrets other than just skeletal structure emerging from fossils

As well as helping palaeontologists to answer important questions about whether or not dinosaurs were warm-blooded, their fossil remains are beginning to relinquish other scientific secrets that had previously been overlooked.

In one extraordinary recent example, Chinese and American researchers managed to reconstruct from fossil remains the real pigmented appearance of a bird-like, feathered dinosaur known as a 'troodontid', which lived about 150 million years ago during the late Jurassic period. By putting the fossilised remains of the ancient animal under an electron microscope, Quanguo Li, from the Beijing Museum of Natural History,* was able to pick out the impressions of minute

* *Science* 12 March 2010: Vol. 327, no. 5971, pp. 1369–72
  DOI: 10.1126/science.1186290

structures, measuring just one-thousandth of a millimetre across, called melanosomes.

These tiny bodies are also found in the feathers of modern birds and contain the pigment melanin, the same stuff that colours human hair and skin. Different coloured melanosomes are slightly different shapes, so by looking at the colours of similarly structured melanosomes in birds around now, Li was able to work out the likely colour scheme for each of the feathered regions of the troodontid dinosaur. Consequently, scientists can now paint a truly accurate picture of what this beast would have looked like: a dark grey body with a reddish-speckled face, reddish crown and long white legs with distal black spangles.

Another major palaeontological leap forward in recent years has been the announcement by scientists that they have also managed to extract genuine *T. rex* tissue from a fossil, casting doubt on the claim that fossils are just reptile-shaped lumps of rock. This discovery was stumbled upon by North Carolina State

University scientist Mary Schweitzer[*] in what she described initially as 'a lucky accident'. She had placed fossil fragments into a solution to remove some of the minerals, but forgot about them and left the samples soaking for far longer than she had intended. But rather than dissolving like a tooth in cola, the samples yielded something far more exciting: once the minerals had gone, what remained was a matrix of fibrous material resembling the connective tissue scaffold that binds bone together.

To find out whether it really was dinosaur tissue, Schweitzer and her colleagues used electron microscopy when looking for the characteristic stripy pattern of collagen, one of the major bone connective tissue proteins. Having spotted what they were looking for, they then used antibodies programmed to recognise collagen, and these bound to the tissue too. Finally, to confirm that it really was collagen, the team used a mass spectrometer

*   *Science* 13 April 2007: Vol. 316, no. 5822, pp. 277–80
    DOI: 10.1126/science.1138709

to work out the chemical sequence of the amino-acid building blocks from which proteins are made. When they compared the protein sequences to modern animals, they found a close match with chickens, frogs and newts, some of the dinosaurs' closest living descendants.

'This similarity to chicken is definitely what we would expect, given the relationship between modern birds and dinosaurs,' says Schweitzer. 'From a palaeo standpoint, sequence data really is the nail in the coffin that confirms the preservation of these tissues.'

This suggests that fossils may be more than just a rocky replacement for the real thing. In fact, there may indeed be parts of the real thing still lurking inside. That said, some scientists have publicly expressed doubts about Mary Schweitzer's results, refusing to believe that tissues can survive intact for such a long time. On this occasion, however, this is probably one instance where time really will tell!

Testosterone beefs up money markets?

Ask a financier what decides whether the stock market makes money on any given day and they'll invariably talk about bulls and bears, falling FTSEs, downturned Dows, a nifty Nikkei and possibly the prospect of a double-dip depression. But unfortunate as it is to burst this fiscal bubble, recent research has shown that, rather like Russian female athletes in the 1970s, the performance of the stock market is actually heavily dependent on male hormones.

John Coates and Joe Herbert, two researchers at Cambridge University,[*] followed 17 male City traders in London over an eight-day period, logging their daily profits and losses and comparing their financial performances with their testosterone levels, which were measured using saliva samples collected at 11 am and 4 pm each day.

'We were following a hunch I had when I was working on Wall Street during the dot-com

*   *PNAS* 22 April 2008: Vol. 105, no. 16, pp. 6167–72
    DOI: 10.1073/pnas.0704025105

bubble,' explains Coates. 'I was struck by the fact that the traders at the time were acting very differently before the bubble and after the bubble. [Before,] they were displaying classic symptoms of mania. They were overconfident. They had racing thoughts, diminished need for sleep and they were carrying themselves in such an odd way I began to suspect there was a chemical involved.'

Sure enough, the results from the London traders showed that on days when their 4 pm financial results showed an above-average return, their morning testosterone levels were 25% higher than on days when they made no returns, or lost money.

The researchers had also predicted that on loss-making days the traders would show higher levels of the body's main stress hormone, cortisol. In fact, the saliva samples showed no such relationship. Instead, there was a strong correlation between cortisol and the degree of unpredictability or volatility seen in the market. The harder the market was to read, the higher a trader's cortisol level, which could, Coates thinks, have long-term impacts on the market.

'In terms of affecting traders' decisions, what it can do is affect the memories you recall. You tend to recall bad memories, negative precedents. You tend to see risk where maybe there is none. You become fearful, you feel anxiety. I think that decreases a trader's appetite for risk. While testosterone is causing people to take too much risk, cortisol is causing people to take too little risk in the crash,' says Coates.

So these findings could explain the basis of the boom-and-bust nature of financial markets. When times are good, testosterone levels shoot up and boost confidence, which increases impulsivity and promotes willingness to gamble, inflating financial bubbles. It also provokes the release of the brain's pleasure chemical dopamine, possibly explaining why traders talk about getting a buzz from their jobs. Cortisol, on the other hand, makes traders risk-averse, causing crashes to become entrenched and turning 'dot com' into 'dot bomb'. But can this new knowledge help to dig the world out of the present downturn?

'Cortisol is a hormone that responds not just to loss or injury (loss being in this case money); it also responds more powerfully to situations of novelty, uncertainty and uncontrollability. Within banks, I think it's extremely important to create an environment that minimises the trader's feeling of uncontrollability. Managers think they have to be proactive to show that they're doing something to improve the situation. But usually what they're doing is threatening to fire people. That's exactly the wrong thing you should be doing.'

One also has to wonder what would happen if there were a few more female stockbrokers. Having less than 10% of the testosterone of a male, it's entirely possible that women might not succumb to the same hormone-fuelled episodes of financial overexuberance . . .

Sugar and spice
and all things nice
... and E. coli

If girls are made of sugar and spice and all things nice, then why are their hands covered in *E. coli*? The view amongst the masses (or maybe that should be the great unwashed?) is that women are generally better-scrubbed than men. But this turns out to be a microbiological myth, at least in many instances. The grimy truth, revealed by a study of commuters' hands spanning the length and breadth of Britain, is that up to 30% of women in some parts of the country have hands covered in faecal bugs!

Called the 'Dirty Hands Study', this national quest in pursuit of hand-borne faecal filth was set up by London School of Hygiene and Tropical Medicine researcher Val Curtis and her colleagues[*] in recognition of the launch of Global Handwashing Day in November 2008. Between August and September of the same year, they sent

*   *Epidemiology and Infection* March 2010: Vol. 138, no. 3, pp. 409–14

out students armed with swabs to prowl amongst lines of queuing commuters at the transport terminals of five major cities: London, Cardiff, Birmingham, Liverpool and Newcastle. In total, over 400 commuters consented to have their hands swabbed down to see what was lurking on the skin. The samples were sent to the microbiology laboratory for culture and identification. Nearly 30% of them tested positive for the presence of faecal bacteria, including *Enterococcus* and *E. coli*, two common gut bugs. Yuck!

So who shouldn't you shake hands with while en route? In general, bus passengers and people living in the north of the country emerged as the worst culprits. On average, one in every three bus users was ferrying faecal organisms on their fingers, compared with the approximately one-quarter of rail commuters who were found to be carrying more than just their luggage.

There was also an association between job description and the risk of being found to be foul-handed. Manual workers were amongst the least likely to test positive, growing bugs from their swabs only about 10% of the time, and professionals came up red-handed in about a quarter of cases. But the real typhoid Marys

Stripping Down Science

– with a whopping 40% contamination rate – turned out to be those describing themselves as administrators. Administering what exactly, you may well ask?!

The most interesting results emerged, though, when the researchers divided up the results by sex. In London, 13% of people in total were positive, with women coming up trumps 20% of the time, compared with just 6% of men. But this trend reversed the further north the researchers looked. In Birmingham, mid-way up the country, the rates were about equal between men and women, with about 25% of people testing positive; but in Newcastle, that number was close to a gross 50% of all travellers. Of the women, 30% tested positive along with a massive 60% of the men. The reasons for these trends aren't altogether clear. One possibility is that further north, commuters may have further to travel and therefore the chances of picking up bugs could rise in proportion to the length of the journey.

The bugs that were detected in the study – commonly *Enterococci* – aren't themselves serious pathogens, at least in a healthy person, but they do serve as a useful marker of faecal contamination and general poor hygiene. Other agents that spread similarly – like the dreaded norovirus, which causes projectile vomiting and diarrhoea – could easily be there too (the researchers didn't look) and are much nastier. This is probably the reason why so many people – in the UK, an estimated one in five – experience at least one case of diarrhoea or vomiting each year.

According to Val Curtis, 'We were absolutely flabbergasted so many people had faecal bugs on their hands. If these people had been suffering from a diarrhoeal disease, the potential for it to be passed round would be greatly increased by their failure to wash their hands after going to the toilet . . .'

So were all the people who tested positive practitioners of poor hygiene? Probably not. In all likelihood, many of them had picked up the bugs from surfaces they had touched in the course of making their journeys. All it takes is one person who doesn't wash their hands to touch a surface and whatever's on their skin can then end up on the hands of anyone else who touches that same surface later.

This is why the rules regarding doors on public toilets should be rewritten to ensure that they can be pushed open with feet or a shoulder on the way out, so there's only a need to pull them open with a handle on the way in. People are much more likely to be contaminated when they leave than when they come in, so this strategy would avoid the one person who doesn't wash their hands from upsetting the applecarts, and the stomachs, of the more conscientious handwashers among us.

FACT BOX

Another reason why your hands might land you in it

Apart from criminal levels of contamination on the hands of British commuters, scientists have found that the unique collections of bacteria we leave behind on the things we touch could also be used to place us at the scenes of crimes in future.

Scientists have known for a long time that the human body plays host to a massive community of microbes. In fact, some have gone as far as to describe us as passengers in our own bodies, because the number of bacterial cells living on us and in us outnumber our own cells by at least 50 to 1.

Most of these bugs are so-called 'good bacteria'. In the intestines, they lend their genetic know-how to assist in the breakdown of what would otherwise be indigestible foodstuffs. They also synthesise certain vitamins and micronutrients, and they protect the host from pathogenic microbes by preventing bacterial and fungal 'nasties' from gaining a nutritional toehold. On the skin, they do similar jobs.

But in recent years, scientists have discovered, largely thanks to the power of modern genetic sequencing techniques, that the diverse community of bacteria that we all carry is as unique to each of us as our own fingerprints. In essence, we have a unique and stable 'bacterial fingerprint' (which can also include anything we pick up transiently from bus passengers and people from the north of England). So even after a handwash with soap, within a few hours the same spectrum of microbes, unique to each of us, is back in place on the skin.

Now researchers have shown that this bacterial badge can be used forensically. University of Boulder, Colorado, scientist Noah Fierer and his colleagues[*] took swabs from computer mice and keyboards and were able to produce a genetic profile of the bacteria present and then pick out correctly – from more than 250 possible pairs of hands – the correct computer owners.

Although they point out that further investigation and validation of the technique will be necessary to establish its forensic credentials, the researchers think that it could be a powerful add-on to existing methods, like human DNA fingerprinting. 'This approach might represent a valuable alternative to more standard techniques,' they point out. 'Given the abundance of bacterial cells on the skin surface and on shed epidermal cells, it may be easier to recover bacterial DNA than human

[*]   www.pnas.org/content/early/2010/03/01/1000162107
      DOI: 10.1073/pnas.1000162107

DNA from touched surfaces.'

So, would-be criminals are advised to take a shower immediately before breaking and entering!

LUNCH BOX

Food for thought

Apart from gut microbial fingerprints, scientists have also shocked the world recently with the discovery that one common strain of human gut bug appears to have borrowed a set of genes from a marine microbial cousin – to help it to digest its host's diet of sushi!

Gurvan Michel, a researcher at the University of Victoria in British Columbia,* had been studying a sea-dwelling bacterium called *Zobellia galactanivorans*, which is a member of the *Bacteroides* family of bacteria.

* _Nature_ 8 April 2010: Vol. 464, pp. 908–12 DOI: 10.1038/nature08937

He found that these bugs carry genes that enable them to make a newly identified class of enzymes called porphyranases. These can break down sugar-based polymers known as porphyrans that are present in large amounts in certain types of seaweeds, including the ones used to wrap up sushi.

Bacteria don't generally have a taste for sushi, but they do degrade seaweeds, so Michel searched the international DNA databases looking to see whether any related genes had been found by other scientists elsewhere. Incredibly, similar gene sequences had been seen before – but not where anyone was expecting: the results he turned up were from bacterial samples collected from human intestines. In this case the organism harbouring them was part of the normal human flora and known as *Bacteroides plebeius*.

More intriguing was that the samples in which these bacteria had been identified were all from Japanese people – the genes appeared to be missing amongst the microbes living in the guts of people in America. This suggests that the Japanese liking for seaweed (the average daily consumption is 14.2 grams per person) has at some point carried the DNA from a marine bacterium with seaweed-digesting genetic know-how, into the guts of a sushi-eater (or similar), whereupon the resident bacterial flora have grabbed the genes and incorporated them into their own metabolic repertoires.

In this way, the bugs have equipped themselves with the chemical equivalent of a sharper knife and fork to help them to deal with a tough new foodstuff. As Stanford scientist Justin Sonnenburg, commenting on the research, puts it, 'Next time you take a bite of an unfamiliar food, think about the microbial inhabitants you may be ingesting, and the possibility that you will be providing one of your 10 trillion closest friends with a new set of utensils . . .'

ALSO BY THE NAKED SCIENTIST

The Return of the Naked Scientist
Dr Christopher Smith

More Scientific Secrets of Everyday Life Laid Bare

Why use expensive beauty products when you can moisturise with jellyfish? Have you ever suspected pollution was to blame for your children's plummeting IQ? Ready to take a sea change . . . on Mars?

Science does not sit still and, continuing on from the success of Chris Smith's first book, The Naked Scientist, comes the equally compelling and curious follow-up. In The Return of the Naked Scientist, you will discover a treasure trove of cutting-edge research, far-flung factoids and the ability to see into our scientific future

Available at all good retailers.

214-7
258-61